The Culture of Science:
Is Social Science Science?

Joseph Woelfel
University at Buffalo

RAH Press, Buffalo, NY

The Culture of Science

Joseph Woelfel / RAH Press

Copyright 2013, 2015

ISBN # 9780989269346
ASIN # B00CLV07XQ

Kindle Edition	2013
iBook Edition	2013
Print Edition	2015

Table of Contents

List of Tables

Table 1: Distances from Biff to attributes

Table 2: Distances from Ray to attributes

Table 3: Order of Presentation for three groups

Table 4: Size of Moon, Sun, Nickel and Quarter for three groups

Table 5: Ratios of sizes of Moon, Sun, Nickel and Quarter for three groups

List of Figures

The Culture of Science

Physical concepts are free creations of the human mind, and are not, however it may seem, uniquely determined by the external world. In our endeavour to understand reality we are somewhat like a man trying to understand the mechanism of a closed watch. He sees the face and the moving hands, even hears its ticking, but he has no way of opening the case. If he is ingenious he may form some picture of a mechanism which could be responsible for all the things he observes, but he may never be quite sure his picture is the only one which could explain his observations. He will never be able to compare his picture with the real mechanism and he cannot even imagine the possibility or the meaning of such a comparison.

— **Albert Einstein**
Albert Einstein and Leopold Infeld, *The Evolution of Physics* (1938), 33.

Preface

A good many times I have been present at gatherings of people who, by the standards of the traditional culture, are thought highly educated and who have with considerable gusto been expressing their incredulity at the illiteracy of scientists. Once or twice I have been provoked and have asked the company how many of them could describe the Second Law of Thermodynamics. The response was cold: it was also negative. Yet I was asking something which is the scientific equivalent of: Have you read a work of Shakespeare's?

I now believe that if I had asked an even simpler question — such as, What do you mean by mass, or acceleration, which is the scientific equivalent of saying, Can you read? — not more than one in ten of the highly educated would have felt that I was speaking the same language. So the great edifice of modern physics goes up, and the majority of the cleverest people in the western world have about as much insight into it as their Neolithic ancestors would have had (Snow 1998).
-- C. P.

Science does not progress steadily forward, but

makes mistakes, branches, retreats and starts again, and, through relentless observation, communication and checking, corrects its errors and moves on. Since the advent of modern science, we have passed through numerous serious errors, many of which have been discovered and left behind. Only a few of the most notorious errors are the luminiferous aether, phlogiston, Lysenkoism, Lamarckism, the geocentric universe, Ptolemy's solar model, the brain as radiator, nerves as hollow tubes conducting spirit, stars as fires, stars as ice, Atlas on a turtle holding up the world, the sun as a huge fire on a chariot circling the earth, the indivisibility of atoms, the idea that objects seek their proper place and have their own proper motion, the spontaneous generation of flies, the perfection of the heavenly spheres, the flat earth hypothesis, the elemental particle theory (i.e., that all things are made from earth, air, fire, water), and Aristotle's dynamic theory of motion, among many, many more.

Social science, too, has generated a plethora of overlapping and contradictory theories of human thought and behavior, such as Aristotle's rational actor model, the free will model, the Calvinist predestination model, the uses and gratifications approach, the Freudian psychodynamic model, Jung's theory of cultural archetypes, cognitive dissonance theory, the Wisconsin status attainment model, Marxism, Capitalism, Weber's Ideal Type model, the Galileo model, symbolic interaction theory, Lewin's field theory, Pareto's circulation of the elite, Durkheim's theory of the collective consciousness, Parson's AGIL model, the damped harmonic oscillator model, stimulus-response theory, contiguity of response theory, labeling theory and many, many more. Since the advent of modern social science, the number of these competing theories that have been decisively rejected is -- none.

Why is it that the physical sciences have been so successful at culling unsuccessful theories, while the social sciences have failed to eliminate *any* of their competing theories? Are the social sciences, as Marvin Gardner once said of psychology, a fast race around a short, round track?

The conventional answer, of course, is that the social sciences are much more difficult than the physical sciences due to the inherent nature of the subject matter. Physical phenomena are concrete, specific and easily observable; social and cognitive phenomena are inherently vague and evanescent, spontaneous and not governed by natural "laws." Scientists of any kind, however -- particularly those who understand that the result of observation is not independent of the way observations are carried out -- ought to be deeply suspicious of words like "inherent."

The notion that the subject matter of social science is inherently different from that of physical science, that we know this independent of any observations or measurements, and that such differences require radically different concepts of measurement is an extraordinary claim, and, as such, ought to require extraordinary evidence. Yet the serious scholar will look in vain for any evidence whatever, for this notion is not a scientific finding, but a philosophical belief. Like

Sumner's folkways and mores, such assumptions bind not because they are strongly supported by evidence, but because they are unquestioned.[1]

This is a book about two cultures. The first of these, the Ionian Scientists, can be traced to Thales in Miletus (ca 620 B.C.E. - ca 546 B.C.E.). This is the culture, passed down through a social network which, although sometimes severely diminished, oppressed by the majority culture and often forced underground, embodies the core beliefs which underlie modern science. When Hawking and Mlodonov describe the Ionians in their book *The Grand Design*,[2] and when Schrödinger retraces the origins of science back to the Ionians in his lectures at University College in Dublin,[3] it is clear that this is not just a culture they understand and appreciate, but to which they belong.

The second culture, that of the Athenian Philosophers, has its origins in the objections of Heraclitus and Parmenides to the teachings of the Ionians, and its formal statement in the Athenian philosophers Socrates, Plato and Aristotle. This culture, passed down through a social network that includes Augustine and Aquinas, constitutes the core beliefs that underlie Christianity, and through Christianity, Western culture in general.

This book is part of my lifelong effort to learn what "science" is. As a sociologist, I was trained in the methods of social science, and, like virtually all of my fellow social scientists, my training in the physical sciences and mathematics was inadequate. I was educated by Jesuits both in high school and college. At

Canisius College, we specialized in debate. Once, in a tournament at the University of Buffalo, the UB debate coach, Mrs. Potter, critiqued us by saying: "You Canisius boys are all alike -- you think like Aristotle and speak like Clarence Darrow." My partner and I were baffled: How could that be bad?

In the sociology doctoral program at the University of Wisconsin, we learned what science was and how to do it from Carl Hempel, Paul Oppenheim, Ernest Nagel, Karl , Karl Popper, R. A. Fisher, S. S. Stevens, Rensis Likert and the like. Completely absent from our curriculum were Galileo, Newton, Einstein, Bohr, Born, Heisenberg, Hertz, Poincare, Lorentz, Schrödinger, Jammer, Reichenbach, Schlicht, and other scientists and scientifically trained philosophers who took the time to inquire self-reflexively about what they did.

Reading these and other scientifically trained authors like Feynman and Hawking made it clear to me that the difference between physical and social science went far beyond the object of our inquiry. Over the course of my life as a social scientist, it has become obvious to me that what we have been trained to do is very different from what physical scientists do. As a sociologist, I began to see that these were, indeed, two different cultures embedded in two different social networks that overlapped only incompletely.

I also realized that there is no way this story can be told in an impersonal way. Time and time again my understanding of the culture of social science is based on my own personal experiences as a social scientist trying to understand science, and leaving these out creates gaps that can't be overcome, at least by me. I'm comforted that many of my models -- the people I've relied on to instruct me about what science is, have written deeply personal stories about their own struggle to understand science -- people like Jacob Bronowski, Erwin Schrödinger, Richard Feynman, Albert Einstein and many others.

The Two Cultures

Of course the notion that the Natural Sciences represent a different culture from the rest of academia is not a new idea. In the middle of the 20th century, C. P. Snow[4] suggested that the culture of Natural Science was a new culture born of the Scientific Revolution in conflict with the old culture of academia as represented by the literary elite. His tone is polemic, strongly favoring the scientific culture and arguing that its adoption is essential to the future of humanity, and it brought forth fierce reactions from representatives of the literary culture, particularly F. R. Leavis.[5] As Collini points out in his introduction to the 1998 edition of Snow's *Two Cultures*, this debate is only the most recent in a series of struggles between scientists and literati: "...the Romantic versus the Utilitarian, Coleridge versus Bentham, Arnold versus Huxley, and other less celebrated examples."[6]

The present book revisits the notion of the two cultures, but its purpose is not to join the debate on one side or the other.

Although Snow's lecture is often generalized to refer to a rift between two cultures worldwide, in fact Snow had in mind two specific groups of scientists and literati of his acquaintance. Although he hints at a more general notion, Snow does not make a careful examination of the essential differences between the groups, nor does he undertake any systematic study in support of his argument. In fact, the two cultures weren't really the focus of Snow's lecture at all; his main argument was that the English educational system in particular and the Western educational system in general needed to be changed to produce more engineers and scientists who would be needed to help industrialize the pre-industrial nations and eliminate the gap between the rich and the poor.

Collini argues, with some justification, that Snow has not identified an essential characteristic that differentiates the two cultures, and further suggests that the "sprouting of ever more specialized sub-disciplines and the growth of various forms of inter-disciplinary endeavor" further erode Snow's distinction between two cultures. He understands, however, that these objections don't rule out "...the possibility of there still being something distinctive shared by those activities which are referred to as 'the sciences', and not characteristic of those designated 'the humanities.'" Collini is pessimistic about the likelihood of finding such criteria, since "Philosophers such as Wilhelm Dilthey in the late-nineteenth century or Karl Popper in the mid-twentieth endeavored to draft the relevant conceptual legislation, stipulating the general properties needing to be possessed by a form of enquiry before it could legitimately be designated as 'scientific.' However, none of these attempts has ever commanded general assent, least of all among the philosophers of science."[6]

Still, the notion of a disconnect between physical scientists and the rest of the disciplines is not so easily dismissed. A few years before Snow's lecture, no less a scientist than Erwin Schrödinger, faced with the difficulties of quantum physics, decided to examine the roots of science among early Greek philosophers in hope of finding the origin of a conceptual error that was blinding physicists to what they might otherwise see clearly. So strongly did he feel the cloistering walls of science around him that he felt the "...urgent need of prefacing [my lectures] with ample explanation and excuses.... I did feel a little uneasy, particularly since those lectures arose from my official duty as a professor of theoretical physics."[3] He goes on to spend several pages describing in particular the rift between science on the one hand and religion and metaphysics on the other, suggesting that, if a scientist were to consider any moral or philosophical issue "...he is liable to have his fingers rapped..."[3]

Five years later, Jacob Bronowski, a man who, like Snow, spanned both scientific and literary circles, delivered a series of lectures at MIT in which he describes what he calls an "absurd division" between scientists and "...those whose education and perhaps tastes have confined themselves to the humanities..."[7] Fifteen years later, Bronowski returned to the same theme in the Silliman lectures at Yale University.[8]

The purpose of this book is to explore the culture of science, and to contrast it with the culture of what we now call the social sciences. Collini's pessimism notwithstanding, are there cultural differences between science and society in general, and the social sciences in particular? What do we mean by "culture?" Since culture abides in an underlying social structure, are there differences in the social networks that underlie science and the other disciplines? Specifically, are social science and physical science different cultures that inhere in different social networks? Are the results or lack of results of scientific and social scientific research attributable to cultural differences, or are they simply due to differences in subject matter?

This book is an attempt to show that the scattered condition of social science theory cannot be attributed to inherent characteristics of social phenomena, but rather is a consequence of the underlying philosophy of the social sciences and, as a consequence of this philosophy, the methods of study that they employ. The book further argues that the philosophy and methods are part of the model of reality generated by the social network to which social scientists belong.

Further, this book will argue that it is not individual minds, or brains, that generate models of reality, but rather that these are generated by social networks, and that the model inheres in the substrate made up of the sets of interconnected brains in the social network. The book will also try to show that the social networks of social scientists and physical scientists are distinct, and have been so since at least the third century B.C.E. Moreover, the models of reality developed by each of these two social networks are different, and incompatible.

This scrutiny may seem to ask another question that has already been answered: what is science? In spite of the facile definitions of the schoolbooks, and in spite of the overwhelming success of the past 400 years, scientists of the first rank are not unanimous in their understanding of what they do. Moreover, to assume that members of a culture understand the culture and carry out its mandates day to day in a thoughtful and calculated way is to misunderstand culture. Members of the scientific culture behave in ways that result in "scientific" knowledge, but that does not mean that they are consciously aware of the forces directing their behavior. A good example is one of the most important scientists of the present day, Steven Hawking. In his 2001 book *The Universe in a Nutshell*, Hawking, a proponent of "modeled reality" as the fundamental method of science, says he is a positivist like Karl Popper,[9] but neither Hawking nor Popper -- a major critic of positivism -- are positivists. One of the main purposes of this book is to make explicit what it is about the culture of science that produces this result. And my ultimate goal is to try to construct a social science that physical scientists would recognize as science.

The book identifies two social networks, each supporting a very different culture from the other. The first, the culture of Ionian Science, has certain characteristics which produce science. First and foremost is the nature of knowledge and understanding.

For the Ionians, knowledge consists of constructing agreed upon symbolic models of what happens in the phenomenal world. This description always consists of a symbolic model of the processes described. The model is understood to be a model of no ontological significance, and it is readily replaced by a model that fits more closely to observations, or fits as well but is simpler. Observations consist of comparisons to some standard, and are always comparative rather than categorical. The world is not considered special, exists for no particular purpose, and was not created by any superior beings who need to be worshipped. Ionians don't explain why anything happens, but simply describe what happens. Knowledge is not a perfect, abstract and unchanging thing, but rather is always approximate, uncertain and subject to revision in light of new observations. It is the set of observations about which scientists agree. Finally, this agreement is never complete or perfect, but always approximate to within a certain tolerance, which science strives to reduce. This view of knowledge is consistent with the meaning of the Greek term *doxa*.

These characteristics stand in stark contrast to the Culture of the Athenian Philosophers. For them, knowledge is abstract, perfect and unchanging, as is conveyed by Plato's term *episteme*. Since the world of experience is in constant flux, it is not the object of knowledge. For Plato, of course, the object of knowledge is the ideas from the world of ideas, while for Aristotle the object of knowledge is substantial form, perfect, unchanging forms which are abstracted from experience by induction, and have their ultimate source in the *uncaused cause* or *unmoved mover*. Knowledge is certain. Things have to be "explained," which means deduced from first principles, since Aristotle's goal is perfect knowledge through causes. The world must have a creator, since, by logic alone, we must reject the idea of a dependent chain of being that depends on nothing. Moreover, everything that happens happens for an end, since the cause of any event must precede that event, and the goal or "final cause" is what motivates motion and change; nature is teleological. Lastly, the Athenian model consists of not one, but two worlds -- one, the imperfect, changing, corrupt world that we live in day to day, and another perfect, unchanging world, which is the true object of knowledge. For Plato, it's the world of ideas, for Aristotle, the heavens, for Christians, Heaven, and for social scientists, the population or universe. These other worlds are unobservable, but we may, perhaps, know something of them through dialectic, or induction, or divine revelation, or good sampling and tests against the null hypothesis.

Although many people believe that the two cultures have patched up the differences between them that led the Church to arrest Galileo for saying that the earth orbits around the sun, in fact there are fundamental incompatibilities between them that render mutual understanding impossible, not the least of which is that they have different definitions of "understanding."

This book asks if social science is indeed "science,"

a question which, on the face of it, seems as if it must have been settled long ago. Indeed, when I delivered an earlier version of these thoughts at a University Lecture at the State University of New York at Buffalo entitled "Is Social Science Science?" one social scientist wrote to the university newspaper to announce that he was insulted by my question. But as I write these words, scientists and engineers have landed a vehicle called "Curiosity" on the surface of Mars, which is at this time some 160 million miles away, and only a few weeks ago scientists at the Large Hadron Collider at CERN announced that they have established the likelihood that they have observed the Higgs Boson at over 5 sigma -- about one chance in a million of error. Whatever progress the social sciences can claim since the adoption of our current definition of social science, nothing social science has done is remotely as successful as these scientific achievements.

No one doubts that, but the (too) simple answer has always been that our subject matter -- human beings -- is much more difficult and requires very different methods than the study of physical phenomena. Studying human beings is intrinsically harder than studying merely physical things, all of which are plainly visible and easy to measure. But on a deeply personal level, I can't imagine how getting 400 university undergraduates to fill out a questionnaire can be harder than landing a rover on a planet 160 million miles away, or how a particle that only comes into momentary existence when an invisible Higgs energy field is bombarded with 126 billion electron volts is "plainly visible".

The other answer usually given is that human variables are *inherently* unmeasurable. But how do we know this? The implicit answer is that we have tried to study human beings scientifically, and found that that doesn't work. Early on, even as an undergraduate student, I became suspicious of this simple answer, and began searching the literature to find the research that showed scientific study of human beings didn't work. There isn't any. In fact, what research exists using the methods of physical science to study human beings tends to be superb. Early research by Helmholtz, Weber, Fechner and other pioneers uses the same comparative measurement model that underlies all physical measurement, and their levels of precision far exceed today's common practice.

For whatever reasons, social scientists almost unanimously hold the view that human variables cannot be measured in the same way as we measure physical variables. As a result, instead of developing arbitrary but agreed upon standards against which we compare results of observations, we utilize things like categorical scales (e.g., "physics and sociology are very different: 1=strongly disagree, 2= disagree, 3= neutral, no opinion, 4=agree, 5 = strongly agree), or ordinal scales (e.g., Please list your five favorite ice cream flavors from most to least favored), dichotomously (e.g., are you a religious person? Yes or No), and the like. So casual is our respect for precision of measure that we don't even accept precision that is easily obtained -- so, for example, we can easily obtain a very accurate measure of a person's age by asking for their date of birth -- which most members of industrialized nations can give in an instant, but instead we categorize even this into boxes like 25-29, 30-35, 36-39 and so on.

What are the consequences of using different methods of measurement for social and physical sciences? After all, the essence of a good experiment is that subjects under study be treated identically; if the result is different, then the things being studied are different. But if you treat them differently, and the results come out different, you've learned nothing.

That's exactly the situation we have in the case of physical and social phenomena. We examine social and physical objects differently, using different measurement rules and different analysis procedures and we get different results. Yet we say this "proves" that social and physical objects are fundamentally different.

The purpose of this book is to show that the differences in outcomes between physical science research and social science research are not attributable to inherent differences in subject matter, but are a result of the mistaken and flawed methods of social science research -- methods dictated by the culture of social scientists. More specifically, I will try to show that the procedures developed by the Ionian scientists and passed down through a social network dating from Thales to contemporary physical scientists and inhering in the culture of physical scientists is responsible for the success the past 400 years have shown, while the procedures developed by the Athenian philosophers and passed down through a social network including Augustine, Aquinas and Christianity to contemporary social scientists constitute a deeply flawed system inhering in the culture of social scientists and it is these methods that have resulted in the failure of the social science community to make equivalent progress.

1. W. G. Sumner, *Folkways*. (Ginn and Co., Boston, 1906).
2. S. Hawking and L. Mlodinow, *The Grand Design*. (Bantam Books, New York, 2010).
3. E. Schrödinger, *Nature and the Greeks and Science and Humanism*. (Cambridge University Press, Cambridge, 2002).
4. C. P. Snow, *The Two Cultures*. (Cambridge University Press, Cambridge, 1998).
5. F. R. Leavis, Spectator (1962).
6. S. Collini, in *The Two Cultures* (Cambridge University Press, Cambridge, 1998), pp. lxxiii.
7. J. Bronowski, *Science and Human Values*, Revised ed. (Harper & Row, New York, 1965).
8. J. Bronowski, *The Origins of Knowledge and Imagination*. (Yale University Press, New Haven and London, 1978).
9. S. Hawking, *The Universe in a Nutshell*. (Bantam, New York, 2001).

Is Social Science Science?

This book argues that social activities such as science are the product not of individual scientists but of cultures. Cultures inhere in social structures or social networks. Physical or Natural Scientists and Social Scientists constitute two distinct social networks, each with its own culture. These cultures are very different and incompatible.

Chapter 1 begins by defining cultures and social networks in a preliminary way. It describes the formation of a particular network of previously disparate individuals who, as a result of circumstances, found themselves entwined in an emerging culture -- the culture of the University of Wisconsin Dow Demonstration. It shows how a common focus of attention and extensive communication form connections among people and groups of people that define and store patterns of information in a distributed way over the entire emerging network of individuals. It shows that the emerging concept of "The Dow Demonstration" is a collective representation, of which each individual only contains a tiny part, and the whole of which is a property of the entire network.

Chapter 1 also presents a simple experiment that shows how concepts are formed first as collective representations in the set of individuals involved, and only later, through communication, can be distributed to individual people.

Chapter 2 follows Ernst Schrödinger in his quest to discover the origins of modern scientific concepts and methods. He takes us to Ionia in the fifth century B.C.E. to what he believes to be the first scientists: Thales, Anaximander and Anaximenes. First among the concepts on which modern science rests he considers the belief that nature can be understood, and can be understood without reference to supernatural or spiritual forces. Rejecting the literal meaning of these thinkers' beliefs that the world is made of earth, water, air and fire, he rather accepts the notion that the universe is made of one substance, which is transformed into different appearances by a process of compression and rarefaction. This led to the atomic theory of Leucippus and Democritus, which he believes to be the direct ancestor of modern atomic theory. The concept of tiny elementary particles that move closer to each other (compression) or further apart (rarefaction) naturally gives rise to the concept of space through which the particles move.

Chapter 2 also introduces the Ionian method of comparative measurement, where all experiences are measured as comparisons to some standard. Thales, when confronted with the problem of measuring the height of the pyramids, noted that, at a certain time of day, the length of his shadow equaled his height. Thus, at that same time of day, one would only need to measure the length of a pyramid's shadow to determine its height. This method of comparison to some arbitrary standard is the fundamental notion of measurement in science to this day.

Chapter 3 encounters the first opposition to the science of the Ionians in the form of Parmenides of Elea. Parmenides argues that the methods of the Ionians can never arrive at perfect, unchanging, certain knowledge, however useful they may be for practical matters in the everyday world. Moreover, Parmenides rejects the concept of space, which he calls the void or "no thing," since no thing cannot exist because it is no thing. Since space doesn't exist, motion and change are impossible. This marks the beginning of a split between philosophers who seek after perfect, unchanging, certain Truth, and scientists who seek to develop uncertain and tentative but ever more precise models of their experience.

Chapter 4 shows the direct communication link from Parmenides and the Eleatics to Socrates, Plato and Aristotle in Athens. The Athenians are deeply influenced by Parmenides' arguments, and define knowledge (episteme) as perfect, certain, abstract an unchanging. Such knowledge can't have its origin in our ever changing, uncertain and aleatory world, and, in fact, too close a preoccupation with observations only makes one more confused and uncertain. The Athenians reject the tentative, comparative, iterative and observational methods of the Ionians in favor of categorical reasoning and, through Aristotle's work particularly, inductive inference to the essence of things through their forms. Aristotle introduces the notion of the Entelechy -- the idea that everything that happens happens for a purpose. This notion applies to every aspect of the world, including people, so the idea that all human actions are goal oriented -- that every action is intentional -- is a foundation of Athenian belief.

Chapter 5 describes how rapidly growing Christianity adopted the philosophy of Plato and Aristotle along with the teachings of early Christians about the life of Jesus and the ancient Hebrew bible as the foundation of their own Christian philosophy. Christianity was initially hostile to Ionian science and repressed it vigorously, while promulgating its Athenian based philosophy through theology, literature (Dante, Shakespeare) and into modern psychology (Freud).

Meanwhile, Chapter 6 shows the thin thread of Ionian science living on through the work of Copernicus, whose knowledge of the works of the Ionians such as Aristarchus of Samos, and whose use of the Ionian comparative method of measurement led him to believe that not the Earth, but the Sun lay at the center of the solar system. Refinements of these comparative measurements by Tycho Brahe as interpreted by Johannes Kepler led to the rejection of the Platonic idea that the orbits of the planets were perfect circles, but were instead ellipses with the Sun at one focus. Galileo's intensified use of the methods of the Ionians led to great scientific progress, including a description of the law of falling bodies, the discovery of satellites of Jupiter and an ultimate confrontation with the Church. The consolidation of all this Ionian progress brings about Newton's Laws of Motion, and classic Newtonian science.

Copernicus believed that his heliocentric model of the solar system was not just a model, but a true and accurate picture of the world. Many of his contemporaries thought differently, however, and understood it to be simply a useful model that represented observations. Isaac Newton, too, believed that his laws were a true statement of the nature of motion, and that his description of space was a true representation of what was actually there. But Chapter 7 shows how the rigorous application of Ionian comparative measurement wears away the notion of absolute truth inherent in the Athenian model. Increasingly precise measurements leading to relativity theory and quantum electrodynamics spelled the death knell for the idea of absolute truth, and led to the explicit understanding of modeled reality, which is the idea that all theories are symbolic models which represent the pattern of observations, but which have no ontological status other than as invented models.

Chapter 8 summarizes some of the concepts underlying the growth of modern science, and characterizes them as evolutionary development based on the competition for survival of the Ionian and Athenian cultures. It argues that the comparative method of the Ionians leads to the development of the concepts of distance, space, time, velocity, acceleration, force, mass, inertia, energy, power and work. These concepts have no meaning in the categorical model, since they are all based on comparisons or ratios. In contrast, models built on the categorical foundation are static, perfect, unchanging, hierarchical and resistant to development and change.

Chapter 9 focuses on the origins of modern social science. Again it contrasts the approach of the Ionians and the Athenians, this time in terms of their respective understanding of human phenomena. Ionian thinkers like Weber, Fechner, Helmholtz, Mach and others sought a better understanding of the world of experience through a better understanding of human sensory mechanisms, and gradually evolved an understanding of how the senses and the brain created an internal symbolic representation of experience that was not a simple eidetic picture, but rather a symbolic model. Mathematicians like Gauss and Riemann understood that our concepts of space are not *a priori*, but rather based on evolutionary experience, and that other spaces beyond our concept of Euclidean space were possible.

Meanwhile, philosophers and others dedicated to further discussions, particularly of the work of Mach formed a discussion group that later morphed into the Vienna Circle, a group usually defined as "logical positivists." Although the work of Mach and the others was laying the groundwork for the "modeled reality" underlying modern science, the Circle and its followers and critics were still caught up on the idea of finding out if theories were true or false. As logical positivism declined, one of its sharpest critics, Karl Popper, presented his concept of "falsification." If theories can't be proven true, at least false theories can be identified and weeded out, Popper claims. Popper's notion of falsification, along with the statistical ideas of Karl Pearson and his followers would go on to become the core methods of quantitative social science.

Chapter 10 discusses the rise of the disciplines of psychology and sociology. It shows how Helmholtz' assistant, Wilhelm Wundt, became the first person to call himself a psychologist, and form the first psychology laboratory in Germany. Wundt furthered Mach's notion that people formed symbolic models of their experiences that were not simple linear representations of the stimuli. Emile Durkheim came to Germany to study with Wundt, and began to understand the priority of the social system and culture over the individual, arguing that concepts are formed in the collective and only later communicated to the individual.

But Durkheim's work, while greatly influential in sociology, would go unheard by psychologists for the most part, and the next generation of psychologists would know little of the Ionian tradition. The early history of psychology is littered with moral philosophers, Wundt's students who misrepresented the works of their teacher, and scandals. Among the most important influences on the fledgling discipline are Francis Galton and his protégé Karl Pearson. They and their followers developed the correlation coefficient, the standard deviation, and the test against the null hypothesis (TANH) which go on to be at the core of the standard method of social science.

Chapter 11 discusses the correlation coefficient in depth, and shows what it is and how it is calculated. It is, in fact, the cosine of the angle between the vectors that represent two variables. Most important, it contrasts the nature of the correlation coefficient as defined by Karl Pearson and the meaning of "correlation" in physical science. In physical science, theory and observations are correlated when every observation is the same as the theoretical prediction to within measurement error. In Pearson's model, two variables are correlated if there is less than a one in ten or twenty chance that they are perpendicular. Chapter 11 then goes on to illustrate some of the distortions that are introduced into observations by the use of the correlation coefficient.

Chapter 12 examines the earliest concepts of space in the social sciences. Another student of Wundt, Charles Spearman, as were so many of his generation, was taken by the idea that people differed in mental ability, and set out to measure the amount of mental ability people had. In order to discover the "factors" that underlay intelligence, he developed a technique called "Factor Analysis." He was followed in this by L. L. Thurstone, who contributed to the development of the procedure.

Factor analysis produces a space in which the factors are dimensions and the loadings of items on the factors are their coordinates in the space. The space of factor analysis is more restricted than the concepts of Newtonian and especially Einsteinian space of relativity, or the high dimensional, non-Euclidean spaces conceived by Bernhard Riemann, but it was one of the earliest incidences of a concept of space in the social sciences.

Ultimately the idea that individuals' absolute intelligence is determined unequivocally at birth floundered, particularly in the face of evidence from neuroscience that shows the wiring of the brain is not complete at birth but develops in response to environmental factors.

Chapter 13 describes some of the history of how the last piece of the standard quantitative social science methodology -- the categorical scale -- fell into place. It shows how L. L. Thurstone conceived of attitudes as positions in space, and developed a quasi-comparative method for measuring attitudes (his method involved the use of categories in pursuit of a quantitative, comparative measurement device). Thurstone located the individual in the attitude space near the attitude positions with which they agreed. But Rensis Likert, in an effort to improve the scale, developed a device which attempted to measure *degree of agreement* on a five point scale. This scale correlated highly with Thurstone's more tedious scale using Pearson's Product Moment Correlation, and so, in practice, social scientists abandoned Thurstone's more precise method for Likert's much simpler category scale.

Chapter 14 goes on to argue that the combined use of Likert and Likert-type scales, correlation coefficients, standardized data and tests against the null hypothesis is a crude system which offers virtually no power to compare any theory to experience in comparison to the method of the Natural Sciences. It shows that using the methods of the social sciences to address well-known simple physical processes results in very serious distortions. It also shows how the using standard social science methodology on real social data leads to incorrect judgments with real and costly effects on real people.

Chapter 15 takes us back once again to William H. Sewell, the Chancellor of the University of Wisconsin during the Dow Demonstration. Sewell, like many of the most prominent social scientists of the time, was drafted into the US military during World War II. There they worked on interdisciplinary teams to carry out social research on the effects of bombing and other topics. They found the interdisciplinary work stimulating and valuable, and came back from the war to establish interdisciplinary social science programs in a number of leading universities, many of which were under the rubric of social psychology.

Meanwhile, another of Wundt's students, George Herbert Mead, realized that Wundt's theory presupposed the existence of fully developed individuals. But human beings don't arrive into the world fully developed and functioning. Mead realized that individuals were born into a culture that already existed, and developed into the kinds of people they became as a result of interacting with the already existing society. Like Durkheim, Mead's work had little impact on psychology, which was imbued with Aristotle's notion that all the needs and drives were built into the individual as part of its human nature, but was very influential in sociology in general and Sewell in particular. He took a keen professional interest in socialization, the process whereby children learned to be adult members of society.

After some initial successes, he judged the enterprise a failure, particularly in contrast with the explosive success of the physical and biological sciences. Although he believed a number of factors contributed to the lack of success, first and foremost was his belief that social psychology did not develop any major theoretical breakthrough.

In Chapter 16, while Sewell was having his darkest hours in the Chancellor's office in Bascom Hall, a few yards West on Observatory Drive in Agriculture Hall, his student and colleague Archibald O. Haller had organized a study to measure the effects of adolescents' significant others on their educational and occupational aspirations. His team of researchers found that the young people's own aspirations were very close to the average of the expectations of their significant others. The idea that attitudes tended toward the average of expectations led to a theory of attitude change called the Theory of Linear Force Aggregation. This theory was interesting for three reasons: first, it was closer to accounting for observations than competing theories, and second, it was a comparative model that depended on comparative measurements, and third, it implied that the formation and change of attitudes could be modeled as movements in space.

Chapter 17 introduces the Galileo model of cognitive and cultural processes, which is a generalization of the Theory of Linear Force

Aggregation. That theory implied that cognitive processes could be modeled as movements in space, but could do so only crudely. In this chapter we summarize the process by which the Illinois research group worked out how to construct an appropriate mathematical space, and shows some of the very early results.

Chapter 18 describes the early research that showed that Galileo space was both high dimensional and non-Euclidean, as anticipated by Bernhard Riemann a little over 200 years earlier. It also shows a bit of the process by which Galileo researchers were compelled by observations to modify their early concepts of space. This is important to understand, because it marks an important turning point in social science. Before the Galileo model, theories remained essentially immune to rejection or even modification by observations, but the fully comparative Galileo model has to be modified and expanded continually to fit to new observations.

Chapter 19 shows how Galileo space serves as a common frame of reference that makes unambiguous communication among different groups possible. A well known problem in the social sciences since the 20th century is that different individuals and cultures view the world through their own frame of reference, and observations are filtered as a result. Galileo allows observations to be made with a common, conventional method and projected onto a common reference frame so that artifactual differences can be transformed away. More importantly, the concept of transformation to a reference frame in which the mathematical description of process is simplified is one of the most important tools in modern science.

Chapter 20 describes the growing awareness of Galileo researchers of the more holistic character of collective consciousness. Earlier work concentrated on establishing the characteristics of the space, such as its dimensionality and its non-Euclidean character, and dynamics focused on simple messages. In Chapter 20 we begin to see that the collective consciousness is a pattern recognition, storage and retrieval system, and that the Galileo model can be used as a precise measurement device for these patterns. Instead of sending two or three brief messages (such as "Pigs are beneficial and attractive") experimenters began sending complex patterns of information, which the collective can recognize (albeit systematically distorted in some ways), store and retrieve very precisely.

Chapter 21 describes the developing interest in and understanding of neural networks, both real and artificial. This increasing understanding of the neural basis for cognitive and cultural processes provides a foundation for the Galileo model, where we understand that the distances between objects in Galileo space are a function of the degree of connectedness of the neural structures representing those concepts.

Chapter 22 shows how different architectures of simple neural networks can model cognitive processes, and further elaborates the relationship between process in Galileo space and corresponding process in the neural networks underlying them.

Chapter 23 summarizes the results so far, and shows how the use of the comparative method of the Ionians led to the development of the Galileo model, but, more importantly, shows how the use of the Ionian comparative model makes experience observable, and argues that more extensive application of Ionian scientific method to human cognitive and cultural phenomena can bring about a greater self-awareness and understanding of ourselves.

Chapter 1: The Illusion of Reality

And now, we watch the birth of a concept. It is the concept of the Dow Demonstration at the University of Wisconsin in October 1967. We see that it develops first in the collective experiences of dozens, then hundreds, then thousands of people, each of whom has a part -- but only a part-- of the whole.

On October 18, 1967, representatives of Dow Chemical came to the Madison campus of the University of Wisconsin to recruit engineering students. Dow had gained some notoriety among anti-Vietnam war protesters because it was the sole-source manufacturer of napalm, a jellied form of gasoline that clung to human flesh while burning fiercely, and so a small group of students staged a sit-in demonstration at the Commerce Building on campus. Since the Commerce Building was across the street from the sociology department, and since I knew some of the people in the demonstration, a young almost-sociologist (me) went across the street to observe.

I recalled the demonstration being quite small, and the photo above confirms my recollection. There are about 50 people in the photo (the photo is cropped, but the original photo doesn't show many more people). Some, like the young woman with the large bag, appear to be passersby. Others, like myself, are simply watching for whatever might happen, but are not part of the demonstration. A few, like the young woman with the sign, are clearly part of the demonstration. Most of the demonstrators are inside the building in the first floor corridor. Interestingly, none of the original organizers of the demonstration are in the building.

A few of the observers are clearly opposed to the demonstration, and chant "Dow makes soap!" from time to time. From their viewpoint, the demonstrators are "dirty Hippies," although no Hippies appear to be in the photo, dirty or otherwise (I suspect that, in October, 1967, the autumn following the "summer of love," only a small handful of people in Madison, Wisconsin would identify themselves as Hippies.)

A little after noon, a young woman on lunch break from her job at a legal firm stopped by the demonstration. A police officer stopped her on the way in and warned her "You can't go in there -- there are dangerous radicals in there!" She said, "Those people aren't dangerous. They're my friends." She visited a few minutes and went back to work.

Later, 30 City of Madison police with helmets and nightsticks joined the campus police, who were without helmets and sticks because they didn't expect a violent confrontation. These two forces had very different attitudes toward the university and the students, with the Madison police much less sympathetic and perhaps even hostile to what they considered spoiled rich children. Some accounts say Ralph Hansen, the chief of the campus police, asked Chancellor William Sewell to call the Madison police; other accounts say he asked permission to call the police and called them himself. Sewell said in public and privately to close friends that he did not call the police, did not authorize anyone to call the police, and did not know who called them.

According to eyewitnesses, one of the police took up a bullhorn and gave the demonstrators two minutes to leave the building, but only a few seconds later, police began breaking down the glass doors to the building with their nightsticks. As demonstrators exited the building, they were struck repeatedly with nightsticks. As the onlookers and counter demonstrators moved in to see what was happening, the police attacked them as well. 47 students and 18 police officers were hospitalized.

I retell this story here because it is an excellent opportunity to watch a concept being formed -- in this case, of course, the concept of the Dow demonstration.

As this is being written, the Dow demonstration has been over for 45 years, and exists only as a memory. But where is that memory? And clearly, different observers with different perspectives will remember it differently. The concept of the demonstration will be quite different for the demonstrators, passersby, counter demonstrators, campus police, Madison police, sociologist-observers, administrators, the young woman from the law office, Madison residents and others all over the world.

These memories will not only be influenced by the cultural biases of the observers, but will often simply be wrong: while the photograph confirmed my memory that the initial attendance was quite small, my memory that the day was warm and people wore summer clothes is contradicted by the photo of people in coats and scarves. Objective data indicate it was a cold, windy, overcast day, with a maximum temperature of only 53.1 degrees and an average wind speed of 14.27 mph.

Some of the immediate experiences themselves were confused because of the hectic situation: one young woman repeatedly called out "Call the police! Call the Police! Will someone please call the police?" while a young man near her shouted back "Those are the police!"

Some memories, like those of Chancellor Sewell, who died in 2001, are gone.

Of course, we can always say that the concept of the Madison Dow Demonstration exists in the minds of those who experienced it, either directly by being there, or indirectly by hearing or reading about it later, but this will lend nothing to our understanding until we know what a "mind" may be. It is important, however, to understand the plural, collective nature of "minds." While each individual person may have some concept of the Dow demonstration, the totality of the concept spans many minds and is thus not a psychological concept, but a cultural concept. The Dow demonstration is a complex cultural concept whose meaning cannot be expressed in a short definition, nor can it be encompassed within a single "mind." It is, rather, the collective idea belonging not to any person but rather to a social network of people. It is what Emile Durkheim would call a *collective representation*.

In the earliest moments, the concept of the Dow demonstration exists as small fragments in hundreds of individual brains. As participants talk to each other, the bits come to be shared. Reporters interview and record the experiences of individual participants, and individuals can compare their experiences. The concept of the Dow demonstration can be communicated to others who were not present. An individual can listen to these interviews, review film taken at the time and develop a broader and richer concept of the Dow demonstration, but no individual will ever encompass the entire concept. Nonetheless, the collection of all participants, then and later, does contain the entire concept. The concept is formed first not in individual minds, but in the collective. Only afterwards, through communication, can individuals share what the collective already knows. It exists as interconnected patterns of neurons distributed across multiple brains.

Science, and social science, are both also complex cultural concepts or collective representations residing in social networks. No single individual can contain the full meaning of science, or social science. These collective representations reside in a neural network, not a neural network that resides in a single brain, rather a network of brains. They are immensely larger than the concept of the Dow demonstration, they formed over a vastly larger time span, and they inhere in vastly larger social networks, but in principle are the same as the Dow demonstration: distributed representations embedded in changing and evolving social networks of human brains, where only the smallest fragment of the overall concept exist in even the finest brains of the best and brightest among them.

In general, neural networks, of which social networks are an important class, identify, store and retrieve patterns in the ongoing flux of information that makes up the world of experience. The main idea underlying a holistic view of networks is that it does not conceive of the cognitive processes as taking place within the nodes, (or brains, as in the case of the global neural network of brains) but rather within the network taken as a whole: The fundamental hypothesis of this theory is that the structure of cognition at any point in time is given by the collective state of the network underlying that cognition, and *cognitive processes* may be considered a function of the changing state of the network over time.[1]

Neural networks can identify, store and retrieve patterns. These patterns inhere in the network itself, and not in the individual nodes. Figure 1 illustrates this in a simple way:

Figure 1: Flashcards at a stadium

In Figure One, each node (in this case, each node is a spectator at a football game) raises one of a set of numbered cards above his or her head. The result is a pattern that inheres in the entire set of all nodes, but which is not accessible to any single node or subset of nodes. Indeed, in the case of Figure One, a camera has recorded the overall pattern and we can see it, but it is entirely possible and most often the case that no conscious entity is aware of the patterns inhering in a given network at a given time. Any given spectator only needs to know that he or she is holding up card B7, and need not know what the pattern represented by all the cards may be. (The fact that this "photo" may be faked is irrelevant.)

One of the most important products of networks is concepts. Plato thought that concepts originated in the World of Ideas, and what we knew of them was a result of remembering something of what we knew while we were there in a previous life. Aristotle thought that individual people could generate concepts by a process of abstraction from observations of this world. In the late 19th and early 20th century, the French sociologist Emile Durkheim suggested a third possibility -- that concepts were developed in what he called the "collective consciousness." As Woelfel, Danielsen and Yum point out,

> He (Durkheim) suggests that the collective consciousness is the source of concepts: "In [Elementary Forms...] we have tried to demonstrate that concepts, the material of all logical thought, were originally collective representations"[2] A function of the collective consciousness, then, is the formation of concepts. This is at odds with those psychological approaches which consider concept formation to take place in the individual mind...[3]

There are no two snowflakes, trees, bushes or rocks that are the same. Yet over the millennia, the collective consciousness has identified continuities of patterning that it has attached to the terms "snowflake", "tree", "bush" and the like. Durkheim had no idea what kind of neurological mechanism might underlie such a process, but neural network models developed since the latter part of the 20th century provide such a mechanism.

In classic Aristotelian categorization, objects fit within a category if they share the essential attributes that define the category. If a being is *rational* and an *animal*, he is a man, according to Aristotle, since *rational* and *animal* are the essential attributes of man (Aristotle did not believe women were rational).

But in a neural network, nodes become active when impacted by energy. Some research[4] indicates that a rod in the retina can be activated when impacted by approximately 5 quanta ($\sim 10^{-11}$ ergs) at the wavelength where it is maximally sensitive (about 5100 A). Of course, it is never the case that only a single rod is activated, but rather a pattern of rods and cones are simultaneously activated when light from the surround strikes the retina. This pattern of activated rods and cones is communicated along the optic nerve to neurons in the brain. As soon as the eye moves or the scene changes, of course, a different pattern of activations will appear. These patterns are simulacra of the pattern of energy striking the retina, albeit systematically distorted in several ways.

When neurons are simultaneously active, connections begin to form among the active set. As soon as they deactivate, those connections begin to decay. When nodes are so frequently co-active that the growth of connections exceeds the rate of decay, they become linked. The linkages among the nodes are the memory of the recurrent pattern. When exercised, links are strengthened -- they become larger, tighter connections; when they are disused, they shrink. These clusters of interlinked neurons are the concepts by which we represent our experience.

When any subset of the nodes in the pattern becomes activated, they transmit electrochemical signals to the other neurons to which they are connected. Each of these neurons sums up the energy arriving through the connections, and, if the level of activation exceeds a threshold level, the neuron will itself become active. If a sufficient subset of neurons is activated, their activation will communicate through the connections to the remaining nodes in the pattern, which become active, and the entire pattern is activated. In this way a stored pattern can be recalled. Thus, any sufficient subset of nodes in the pattern can serve as a symbol of the pattern -- a trigger to activate the entire pattern.

A category in a neural network, then, is not defined by essential attributes, but by connections among nodes. If the network experiences a pattern that is *similar enough* to the stored pattern, the pattern becomes active. The essential element for identifying any experience, then, is similarity to previously stored experience. If an object is *similar enough* to previously experienced trees, then it is a tree. The pattern for "tree" is not simply stored in individual persons, but in the entire network of people that constitutes the social structure of society. *It is the network that learns new concepts and the relationships among them, not the individual nodes.*

Consider the following experiment: Individual nodes (people) sit at computers and read a paragraph describing six people. The paragraph has been constructed by a random process, whereby people are assigned attributes (e.g., tall, short, intelligent, unintelligent, etc.) by a coin toss:

> Raul is very conservative, intelligent, and very tall.
> Varsha is very conservative, unintelligent, and tall.
> Biff is liberal, very unintelligent, and short.
> Lurlene is liberal, unintelligent, and very short.
> Bobbie is conservative, intelligent, and short.
> Ray is very liberal, intelligent, and very tall.

After reading the paragraph once, the respondents were tested and scored about chance in remembering the paragraph. It is possible to show, however, that the set of nodes as a whole has learned the paragraph. If each individual is asked to estimate the difference or dissimilarity of each fictional person and attribute from every other fictional person and attribute on a comparative scale, the averages of all the responses produce a square matrix of dissimilarities whose principle axes are covariant tensors defining a Riemannian space in which each of the persons and attributes is represented by a contravariant tensor.

Figure 2 is a rendering of the relationships among the fictitious people and the attributes for 80 respondents as perceived by the network as a whole. Like the stadium picture of the creation in Figure 1, none of the individual nodes knows the pattern of relationships among the fictitious people and their attributes, *but the network taken as a whole knows it.* If the individuals talk among themselves at random, they, too, can each learn the pattern, but the network learned it first.

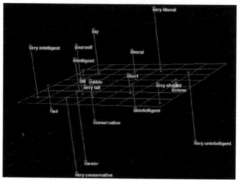

Figure 2: Distances among six persons and their attributes[1]

It is not the case that the network will learn the pattern without distortion. Different networks will selectively distort the patterns to which they are exposed in different ways -- in fact, a main point of this book is to argue that the network in which social scientists find themselves will perceive the world in a very different way than the network of physical scientists.

---

[1] Grid and stems are only to aid in visualizing the distances among the concepts, and have no substantive meaning. Nor does direction; the space is identical if turned upside down or rotated by any angle. So, Varsha is close to "very conservative," far from "liberal," and even further from "very liberal."

In the case of the fictitious people, the network of approximately 80 undergraduate students sometimes gets the pattern right (that is, the person is closer to the attributes assigned in the paragraph than to any other attribute), as it does for Biff,

Biff is liberal, very unintelligent, short.

Attribute	Distance[1]	N
very conservative	149.81	78
conservative	184.00	77
liberal	68.46	76
very liberal	163.81	78
very intelligent	159.33	78
intelligent	194.35	77
unintelligent	112.33	76
very unintelligent	100.57	77
very tall	269.25	77
tall	125.86	77
short	60.86	77
very short	143.12	77

Table 1: Distances from Biff to attributes

but sometime shrinks extremes, making *very liberal* into *liberal* and *very tall* into *tall*, for example, for Ray. While Ray was assigned the attribute very tall in the paragraph, the collective network shows him closer to tall (77.16) than to very tall (85.85). And while the random paragraph said he was very liberal, in the collective network he is closer to liberal (86.71) than to very liberal (152.09).[5]

Ray is very liberal, intelligent, very tall.

Attribute	Distance	N
very conservative	142.97	77
conservative	119.56	75
liberal	86.71	77
very liberal	152.09	77
very intelligent	106.25	77
intelligent	80.63	75
unintelligent	111.36	77
very unintelligent	160.79	77
very tall	85.85	79
tall	77.16	77
short	118.54	78
very short	148.99	77

Table 2: Distances from Ray to attributes

The corpus of human knowledge is stored in books, libraries, records, computer systems, in the architectural styles of its buildings, the form of its art, music and other artifacts. But above all, human knowledge is stored in the immense neural network of human brains. Few investigators, however, have approached the concept of network at this scale. Most work in social network analysis concerns itself with the effects of different kinds of locations in networks of different topologies on the individuals who occupy those statuses and is socio-psychological in focus. Here we will focus on the global network of interconnected brains.

At this scale, individual brains are insignificant and replaceable, each making up about 1.4×10^{-08} of the overall network, and, in fact, are replaced at the rate of about 8% a year worldwide, so that, every century or so, (almost) all individual brains are replaced, yet the cultural patterns stored in the network remain. In fact, the emergent properties of the network itself go far beyond the capabilities of the individual members of the net.

A few writers have considered the global characteristics of networks not for their influence over their human occupants, but for their own intrinsic characteristics and behavior[1, 6-9], yet overwhelmingly social scientists and computer scientists continue to focus their attentions on individual cognitive processes, and consider networks primarily as sources of influence over these individual processes. Why is this?

Neural networks, whether organic or mathematical, share an important feature: patterns detected and stored early in the life of the network serve as the basis onto which later patterns are superimposed, so that the sequence of learning is important. When a child learns about leopards, the leopard-pattern is built on the domestic cat pattern already in place. We understand the relationship between house cats and leopards not by an Aristotelian deductive process, but because their patterns in our brains are largely made up of the same neurons.

Neural networks tend to be conservative. Once trained, patterns exist in the network as connections among nodes. When patterns *similar enough* to the learned pattern are presented to the network, it perceives them as the old pattern. When the old pattern is activated, the connections are strengthened, and the old pattern is reinforced. Only when the network is confronted with patterns too different from any of those already learned to activate the old patterns will the old connections degrade and new ones be formed.

This model is not insubstantial and immaterial. It exists as synaptic connections among neurons in billions of brains. Consider the cultural system founded in Athens and transmitted through the Christian church. There are currently 2.3 billion Christians (33% of the world's population) and a billion and a half Muslims (22%) whose fundamental beliefs are that individuals form the foundation of society, and choose freely between good and evil alternatives based on a calculation of reward and cost. They believe that the world in which they live is only semi-real and largely evil, and must be overcome in order to achieve salvation in the next, real, world. Compare this to the concept of the Wisconsin Dow anti-war protest. Initially, only several hundred people experienced the event, and later news coverage may have spread the concept widely. Although millions of people may have heard a few seconds of news coverage on TV, the total number of neurons and synapses involved in supporting this concept is vastly smaller.

This is important, because our legacy concept of the mind as a non-material spiritual entity has no concept of mass or inertia. Attitudes and beliefs are assumed to be instantly changeable, with no equivalence of energy expended and change experienced. The material notion of the global neural network, however, makes it clear that energy is required to build up synaptic connections, that energy is required to change them, and that the quantity of energy required is proportional to the mass of the system.

This conservative characteristic of patterns learned early on is exacerbated by "thresholding." In the continuing flux that is the world in which we live, it is impossible that any pattern be confronted in exactly the same way twice; as Heraclitus tells us, you can't set foot in the same river twice. But the neural network only needs the similarity of the perceived pattern to match the old pattern to better than a threshold level; if that threshold is exceeded, the network decides that the current pattern is the old pattern, and perceives it as such. Thus if one sees a friend today, he or she may be dressed differently, wear sunglasses, be in a different context, but nonetheless be *similar enough* to activate the familiar pattern, and we see not a novel person, but an old friend.

Networks can change only when information received by the network diverges from the patterns on which it was trained by an amount larger than the threshold; differences smaller than that will simply reactivate the old pattern. This can happen when the network's environment changes, or by encountering other networks. In the case of individuals, these two mechanisms have been called self-reflexive activity and significant other influence.[10] Self-reflexive activity refers to direct sensory experience, while significant other influence refers to information that is mediated by other persons or media.

Both of these processes are deeply dependent on

the culture that is already in place. Symbolic communication from other persons, of course, depends on pre-existing commonly shared symbols, while direct sensory experience is also dependent on observational practices inherited from the culture. Observations of any sort must be encoded into symbols for communication to be possible. The culture is what makes direct experience and communication as we understand it possible.

All this means that enormous concepts such as Ionian science and Athenian/Judeo Christian philosophy/theology, which formed thousands of years ago and evolved over several millennia, have a deep and fundamental influence over what their members are able to perceive, and are profoundly difficult to change.

1. J. Woelfel, Journal of Communication **43** (1), 63-80 (1993).
2. E. Durkheim, *The elementary forms of the religious life.* (G. Allen & Unwin, London, 1968).
3. J. Woelfel, S. Danielsen and J. A. Yum, *Cognitive theory of collective consciousness.* (RAH Press, Amherst, NY, 2009).
4. M. Jammer, *Concepts of Mass in Classical and Modern Physics.* (Dover Publications, Inc., Mineola, 1997).
5. J. Woelfel, S. Danielson and J. A. Yum, *Cognitive theory of collective consciousness.* (RAH Press, Amherst, NY, 2009).

6.   J. Woelfel and W. Richards, *A general theory of intelligent, self referencing networks.* (RAH Press, Amherst, NY, 2009).

7.   J. Woelfel, in *Progress in communication sciences, 12* edited by W. Richards and G. A. Barnett (Norwood, N.J.: Ablex Publishing, 1993), pp. 21-42.

8.   J. Woelfel, *Neural networks: Application of the cognitive revolution to advertising, marketing, and political research.* (RAH Press, Amherst, NY, 2009).

9.   J. Woelfel, *Artificial neural networks for cluster analysis.* (RAH Press, Amherst, NY, 2009).

10.   J. Woelfel and A. Haller, American Sociological Review **36** (1), 74-87 (1971).

Chapter 2: The Ionian Network and the Culture of Science

And now, we follow Erwin Schrödinger in his quest to uncover the origin of science. He finds it in the 6th Century B.C.E., in Ionia, in the work of Thales, Anaximander, Anaximines, Leucippus and Democritus.

This book is about contemporary social science and the reason it has not matched the achievements of physical science. So why then do we begin in the 6th century B.C.E.? The answer is that I believe that the efforts of social scientists are impeded by prejudices that inhere in the culture of social scientists, and that these prejudices have their origin (as far back as we can see) in Ancient Greece. As Bronowski writes,

> ...all our prejudices about the external world tend to be built into the language of science. Then, when somebody shows that the whole thing was nonsense, that we put our prejudices into it, we are always taken aback.[1]

In this I am self-consciously following Schrödinger's example:

It is, then, natural to recall that the thinkers who started to mold modern science did not begin from scratch. Though they had little to borrow from the earlier centuries of our era, they truly revived and continued ancient science and philosophy. From this source, awe-inspiring both by its remoteness in time and by its genuine grandeur, pre-conceived ideas and unwarranted assumptions may have been taken over by the fathers of modern science and would, by their authority, soon be perpetuated...There is not only...the hope of unearthing obliterated wisdom, but also of discovering inveterate error at the source, where it is easier to recognize.[2]

Thales (ca 624-546 B.C.E.) has generally been considered the first Western philosopher and scientist, although recent philosophers like Daniel Graham[3] prefer to begin their analysis with his student Anaximander because nothing of Thales' writings remains. In general, though, the Greek (and indeed modern Western) fascination with the individual hero has led them to attribute great achievements to great personages whether they did them or not. It was common to attribute the accomplishments of the Milesians to Thales, and it is well known that virtually every accomplishment of the Pythagoreans was attributed to Pythagoras himself, even long after his death. Plato, of course, continued in this tradition by attributing much of his own thought to Socrates.

The previous chapter has argued that concepts are

generally not formed in the individual mind, but rather they are cultural accomplishments of social networks and not the accomplishment of individual persons. The idea that concepts, scientific or otherwise, find their origin in the mind of some individual, then gradually diffuse through the culture is exactly backwards; concepts originate as diffuse patterns in a social network which, through communication, can be recognized by individuals who then "discover" them. Thus the concept of relativity theory existed as a pattern in the network of scientists at the end of the 19th Century, some parts in Poincare's brain, some parts in Lorentz' and Fitzgerald's, some parts in Levi Civitas', some parts in Einstein's. By discussions and review of literature -- communication -- all the parts of the pattern were recognized by Einstein and written down, but they were there before he did so.

Historical records show that, far from being the fruits of isolated brilliant individuals like Thales and Pythagoras, the elementary concepts that form the basis of modern science are found widely distributed not only in Greece, but throughout the eastern Mediterranean, North Africa and across the Indus Valley.

Much of the literature on the cultural differences between scientists and others was written in the middle of the 20th century, and was deeply concerned that the education system of England, Europe and the United States was too heavily biased toward language and literature. The educated person would know Greek and perhaps Latin, but probably not a modern language other than his or her own, some mathematics, but little or no science.[2, 4-6] Much of the conceptual structure of Western culture represented in the collective neural network is built out of Greek building blocks.

So much of the culture of the Western world is based on Greek thought that it is not surprising that modern philosophers and historians often display an overwhelming prejudice in favor of the Greeks. John Burnet's very influential *Early Greek Philosophy* is prefaced by this remark: "...it is an adequate description of science to say that it is 'thinking about the world in the Greek way'. That is why science has never existed except among people who came under the influence of Greece".[7]

A clear case in point is Erwin Schrödinger, who sought a solution to theoretical difficulties in physics brought about by experimental results in subatomic physics by harking back to the origins of science. He believed Thales of Miletus to be the first scientist -- the first person to attempt to understand the world in secular terms, with no reference to gods or spiritual forces. Schrödinger considers the idea that the world can be understood without resorting to magic or spiritual factors one of two important concepts on which modern science rests. The second is that "...the 'understander'' (the subject of cognizance)..." is to be excluded from "...the rational world picture that is to be constructed."[2]

This notion, that the world can be understood, and that a model of its main features can be constructed which is independent of the observer, Schrödinger considers the first step toward the development of science. Thales suggested that, despite the diversity of stuff in our experience, all matter has so much in common that, beneath its appearances, it must all be made out of the same thing. That Thales believed that "same thing" was water, Schrödinger thinks to be insignificant, and in any case, thinks it unlikely that we should take Thales' "water" to be literal H_2O. Anaximander, Thales student, denied that the "stuff" the world was made of was any of the ordinary substances of everyday life, and coined a term for this substance, calling it the Boundless.

The third step, according to Schrödinger, is Anaximander's student Amaximines' notion that the underlying substance (Anaximander thought it was air) changed into everything else through a process of rarefaction and condensation. This, according to Schrödinger -- the notion of condensation and rarefaction -- is the intermediate step leading to the atomism of Leucippus and Democritus, which, in turn, leads in an unbroken chain of transmission directly to contemporary atomic theory. The fact that he chose air as the primary substance is also fortuitous. As Schrödinger points out, Democritus would not know the actual composition of air, but his notion that the primary stuff is gaseous is correct: if he had chosen hydrogen, which, of course, had not yet been discovered, he would have been exactly right.

But Schrödinger would have had to go back much farther in time to find the actual origins of the basic concepts of science. Long before we seek models that abstract the essential elements of our observations and allow us to predict and perhaps control them, it is necessary that there be observations at all. Fundamental to all science is the ability to make observations. Without a conceptual structure, the experience of an untrained neural network is cacophonous and aleatory. *We have to learn to observe.*

Archaeological evidence indicates that human ability to observe precisely and scientifically developed much earlier than Thales, at least as far back as the Indus Valley Civilization, which flourished ca. 3300 to ca. 1700 B.C.E. In its major cities, such as Harappa and Mohenjo-Daro, which flourished from ca. 2600 B.C.E. to ca. 1900 B.C.E., observational skills were already well established:

The people of the Indus Civilization achieved great accuracy in measuring length, mass and time. They were among the first to develop a system of uniform weights and measures. Their measurements were extremely precise. Their smallest division, which is marked on an ivory scale found in Lothal, was approximately 1.704 mm, the smallest division ever recorded on a scale of the Bronze Age. Harappan engineers followed the decimal division of measurement for all practical purposes, including the measurement of mass as revealed by their hexahedron weights. Unique Harappan inventions include an instrument which was used to measure whole sections of the horizon and the tidal dock.[8]

Even the atomic theory of Leucippus and Democritus is also found in the Indus Valley. Kanada, in his Vaishesika-Sutra, presents a highly articulated atomic theory as early as the sixth century B.C.E. (although some writers place him as late as the second century B.C.E.) which includes Thales' elements fire, air, water and earth. The location of Miletus and its main occupation as an international trading center makes it clear that there is communication among the east and west, and that these ideas are not the possession of one or two thoughtful individuals, but rather a common cultural property of an existing social network which includes Thales, Leucippus, Democritus and Kanada.

A major purpose of this book is to identify two distinct social networks in ancient Greece, which I will, for convenience, refer to as the Ionian and the Athenian networks. The first of these, the Ionian network, is itself part of a social network that extends back into Egypt, the middle east, and the Indus Valley and forward to our own civilization. This network is the source of the culture of science. The second, the Athenian network, also extends forward to our own civilization, but extends much further and deeper into Western culture, serving indeed as the foundation of our general culture. While a direct communication chain from these networks to contemporary society can be traced, the Greeks were certainly not the sole inventors of these fundamental concepts.

Democritus wrote that the common part of our experience is what we call real, thus anticipating Einstein's similar remark by two and a half millennia. What both men mean is that, through communication, we compare our experiences with the experiences of others. But the direct comparison of experience is, of course, impossible. What is possible is to assign symbols to the observations and communicate the symbols to others. All communication of this sort is symbolic, and the foundation of even the most primitive observations is the assignment of agreed upon symbols to represent our experiences.

The earliest symbols generated by primitive social networks are lost in the historical past, but these probably consisted of gestures[2] and simple verbal utterances. The gestures and sounds become associated with the rest of the pattern of interconnected neurons, so that seeing the gesture or hearing or uttering the sound activates the entire pattern representing the experience. Comparing experiences clearly goes no further than comparing symbols, so the symbols by which experience is represented have a determining effect on the extent to which we can create agreement about our experiences.

Comparisons using categorical symbols are particularly troublesome. If I say "I saw a yellow cat on my porch yesterday" and you say "I saw a yellow cat in my backyard today," we are a long way from guaranteeing they were the same cat, because our experiences of cats may be very different, as are our experiences of yellow -- and there are a lot of yellow cats in the world.

---

[2] The first explicit designation of the gesture as the primitive source of symbolic communication (probably) comes from Wilhelm Wundt. 9 A. Kim, in *Wilhelm Maximilian Wundt, Stanford Encyclopedia of Philosophy*, edited by E. N. Zalta (Stanford, 2008)..)

The Ionians, however, use a special kind of comparative symbolism, in which parts of experience are compared to other parts of experience. The ivory scale discovered in Lothal mentioned above makes it possible, for example, to compare our experiences of length to within a little more than a millimeter. Thales was allegedly the first person to measure the height of the pyramids. He did so by observing that, at a certain time of day in Egypt, his shadow was the same as his height. At that time, he needed only the measure the length of the pyramid's shadow to determine its height.

These comparisons -- the length of something compared to the ivory scale of the Indus Valley, the height of Thales compared to the length of his shadow; the height of the pyramid compared to its shadow, and finally the comparison of Thales and his shadow to the pyramid and its shadow which finds they are the same ratio -- are the fundamental method of observation in science. We learn about the world of experience by comparing our experiences to each other and across multiple observers. As we improve our symbolic models, we are able to compare more experiences in finer detail and achieve broader agreement to increasingly finer tolerances.

Among those who helped establish the notion of comparative measurement as the basis of scientific observation were the Ionians of Samos, about 47 kilometers from Miletus. Among the key figures who lived and studied in Samos were Pythagoras, Philolaos, Archytas, Aristarchus, and, although from Syracuse rather than Samos, nonetheless a part of the Samite network through his friend Conon of Samos, Archimedes.

The first of these, Pythagoras is alleged to have shown that even ratios of the lengths of strings created harmonious sounds. Archytas later solved the problem of producing three harmonious tones by means of dividing a musical interval into three equal steps. Of course the Pythagorean Theorem, which expresses the sum of the squares of the sides of a right triangle as equal to the square of the hypotenuse, is attributed to Pythagoras, but we know now this was also known to the Babylonians and the inhabitants of the Indus valley.

Philolaos is often cited as the author of the idea that the earth, sun and planets orbit around the "central fire", while Archytas is thought to be the first person to replace the central fire with the sun as the center of the solar system. Aristarchus clearly held that the planets circled around the sun, and calculated the sizes of the earth, the moon and the sun, as well as their distances by comparative, approximate methods, such as comparison of the size of the earth's shadow on the mood during eclipses to the size of the moon itself, or by comparing the length of the shadow cast at the bottom of a well in Egypt and Greece at the same time on the same day. All these are examples of comparative measurements, i.e., ratios.

The notion of approximation rather than perfection is well illustrated by a discovery of Archimedes. Archimedes showed that the area of a circle could be calculated to any level of precision by the method of exhaustion by inscribing and circumscribing the circle with regular polygons of increasing numbers of sides. We could begin by circumscribing a triangle around a circle, and inscribing a triangle within the same circle. We can calculate the areas of the triangles, because we know how to do that, and we can then say that the area of the circle must be smaller than the area of the circumscribed triangle, but larger than the area of the inscribed triangle. There will still be a large gap, but, if we replaced the triangles by squares, the discrepancy will be smaller. If we replace the squares by pentagons, the gap will be smaller still, and hexagons will further reduce the gap. This can be continued without limit until any desired level of precision is achieved. This idea -- that measurements can be made as precise as needed but never perfect -- is an essential idea underlying modern science.

This comparative observational model is the basis of Samite observational skills. Most Pythagoreans believed that the Earth revolved around the sun.[10] Copernicus cites Aristarchus in an early draft of his heliocentric model, although the reference is missing from the final, published version.[2] The social upheaval that resulted from Copernicus' publication of his heliocentric theory, resulting in the arrest and trial of Galileo, makes another point about the Ionian cultural attitudes: moving humanity from the center of the universe to an ordinary planet orbiting around an ordinary star represents a massive shift from the spiritual cosmologies of the past. Even now, one of the prejudices that stifles scientific research in the social sciences is the belief that human beings are special and not subject to the laws of nature.

It's worth noting that Schrödinger considers the Ionians, and not the Pythagoreans, the precursors of science, mainly because of the central role of mystery, *a priorism*, and tendency to place reasoning above observation exhibited by the Pythagoreans, but he acknowledges, with some chagrin, that the Pythagoreans progressed farther than the Ionians in the understanding of the solar system:

> It is an amazing fact, not a little disconcerting to the sober scientist of today, that the Pythagoreans with all their prejudices and preconceived ideas about beauty and simplicity made better headway, at any rate in this one important direction, toward an understanding of the structure of the universe -- better than the sober school of Ionian "physiologoi", of whom we shall have to speak presently, and better than the atomists who succeeded them spiritually.[2]

Schrödinger's explanation for the success of the Pythagoreans is central to understanding one of the key things missing from today's social science model. He says the Ionians may have failed to incorporate the cosmology of the Pythagoreans due to their aversion to the mystical aura surrounding the Pythagoreans, but the Pythagoreans achieved great success in spite of this handicap because of their ability to measure precisely.

Disgust at the unfounded, weird phantasies and

the arrogant mysticism of the Pythagoreans may have contributed to cause a clear thinker like Democritus to reject all their teaching that gave the impression of arbitrary, artificial construction. *Yet their power of observation, trained in those early, simple acoustical experiments about vibrating strings, may have enabled them to recognize through the fog of their prejudices, something so near the truth that it served as a good foundation from which the heliocentric view rapidly sprang.* Sad to say, it was equally rapidly discarded under the influence of the Alexandrian school, who believed themselves to be sober scientists, free of prejudice, guided only by facts (p. 51, emphasis added.)

The "power of observation" that Schrödinger mentions can be seen as a result of the use of the comparative method of measurement as opposed to categorical observations and syllogistic reasoning. These early patterns in the Pythagorean network are the Western foundation of the comparative model of measurement (although precise comparative measurement was well known to the engineers of the Indus Valley several thousand years earlier), along with a language precise enough to communicate its findings among scientists -- the calculus, whose foundations lie in the method of approximation of Archimedes.

Precision matters. Neural networks, which are approximate and thresholded, will perceive an incoming pattern as an already known pattern -- even if it is not the same -- if it is "similar enough" to activate the old pattern. With a crude observational model, small differences are invisible, and older patterns are immune to change. But when expressed as comparisons to a standard -- particularly a worldwide standard like the meter and the second -- the differences between what the classical model predicts and what the comparative model observes are too great to be confused with the old pattern.

A good case in point is Aristotle's laws of motion. Aristotle believes that every body intends to move toward its "proper place." The proper place of light objects, such as fire, is at the periphery of the universe, so they move upward. The proper place of heavy objects is at the center of the earth. The heavier the object, the more strongly it desires to be at the center of the earth, so bodies will fall with a velocity proportional to their weight.

This is wrong, and it is not only slightly wrong, as people usually suspect, so the atmosphere can hide the fact that all objects fall at the same rate of acceleration regardless of their mass, so we have to wait for Neil Armstrong to drop a feather and a hammer on the moon to see the difference. It is manifestly wrong, yet Aristotle's laws of motion stood unchallenged for 2,200 years until Galileo exposed them to methods of observation precise enough to make the discrepancy loom large in experience. (The idea that Aristotle's law of falling bodies is manifestly wrong, yet uncontradicted by observation for over two millennia is so discomforting that even as prominent a philosopher of science as Thomas Kuhn trained in physics, is led to believe that the differences between Aristotle's theory and Galileo's model are too subtle to be noticed without advanced scientific apparatus. He is quite wrong, as we will see in Chapter Six.)

An essential feature of the Ionian network, then, is the comparative method of measurement, a procedure known to the engineers of the Indus valley and communicated to us by the Pythagoreans. In fact, the model of experience constructed by the Ionian network can be seen to have these features essential to modern science:

- The belief that the world of experience can and should be understood
- All measurement consists of comparison to some standard
- Every measurement is uncertain, and precise only to within some tolerance
- Knowledge is based on the agreement of observers, and not the private domain of a priestly class
- Agreement applies only within a certain tolerance, and communication is central to observation and agreement
- The world is a natural phenomenon, not created by gods who need to be worshipped.
- Human beings are an ordinary part of that world

But the Ionian model of understanding the world of experience through comparisons to build an uncertain, approximate but increasingly precise and useful model would not stand unchallenged. Other Greeks were to prepare an alternative understanding of the world separate from Ionian science, one which preserved the idea that the earth was a special place created by divine forces; a world in which human beings were very special. As Schrödinger (2002) says

> The Ionian's attitude lived on with the atomists (Leucippus, Democritus, Epicurus, Lucretius) and with the scientific schools of Alexandria though in different ways; for, unhappily, natural philosophy and scientific research had separated in the last three centuries B.C. much as in modern times. After this the scientific outlook gradually died away, when in the first centuries of our era the world became increasingly interested in ethics and in strange brands of metaphysics, and did not care for science. Not before the seventeenth century did the scientific outlook regain momentum (p. 59).

The separation began with Parmenides (early fifth century B.C.E.) and his colleagues in Elea.

1. J. Bronowski, *The Origins of Knowledge and Imagination.* (Yale University Press, New Haven and London, 1978).
2. E. Schrödinger, *Nature and the Greeks and Science and Humanism.* (Cambridge University Press, Cambridge, 2002).
3. D. W. Graham, *Explaining the Cosmos: The Ionian Tradition of Scientific Philosophy.* (Princeton University Press, Princeton, 2006).
4. J. B. Conant, in *The Copernican Revolution* (Harvard University Press, Cambridge, Massachusets and London, England, 1957), pp. xiii - xviii.
5. J. Bronowski, *Science and Human Values*, Revised ed. (Harper & Row, New York, 1965).
6. C. P. Snow, *The Two Cultures.* (Cambridge University Press, Cambridge, 1998).
7. J. Burnet, *Early Greek Philosophy*, 4th ed. (A. and C. Black, London, 1930).
8. Anonymous. http://indianhistory.info/indus.htm
9. A. Kim, in *Wilhelm Maximilian Wundt, Stanford Encyclopedia of Philosophy*, edited by E. N. Zalta (Stanford, 2008).
10. E. M. Antoniadi, Journal of the Royal Aastronomical Society of Canada **34** (1940).

Chapter 3: The Eleatic Challenge and the Birth of Western Philosophy

And now, the Eleatic Philosophers, mainly in the person of Parmenides, object to Ionian science because it yields only relative, uncertain knowledge. He lays down the criteria for perfect, unchanging, certain knowledge, which science can never achieve. He strongly influences Plato, who turns away from science toward philosophy.

It is said that there are two types of people in the world: those that divide everything into two types, and those who don't. Logical Positivism, a 20th century philosophical system aimed at understanding science, divided all possible statements into two types: those that contained an empirical reference to some possible observation, and those which didn't. Some of the adherents of this view held that any statement which had no empirical referent was meaningless, which was itself a statement that had no empirical referent. This logical inconsistency, among other factors, led to the decline of the movement in the second half of the 20th century.

But an understanding of how the collective neural network works shows that the connection of symbols -- words, numbers, sounds, signs and the like -- to sensory perceptions is more complicated than a simple yes or no dichotomy. Every concept, every sensory experience, exists in a network of interconnected concepts and experiences, and every intake of information anywhere in the network of interconnected concepts influences all the concepts, however slightly.

Assume that I see a cat with a sign tied around its neck saying "cat." I now form a connection between the symbol "cat" and the sensations I experience while looking at the cat. If I hear the cat whine, I will form a similar connection between the whining and the cat. If I see a dog with a sign "dog" hung around its neck, I form a connection between the symbol and the experience. But if I now hear the dog whine, my concept of "cat" will change, because "dog" and "cat" are now connected through their common connection to "whine." From now on, anything I see or in any way sense about "dog" will alter my concept of "cat." In this sense, symbols become dissociated from their substrate, that is, the sensory experiences tied to them. My understanding of the word "cat" is no longer exclusively tied to my sensory experiences of cats, but rather takes its meaning from its relationship to all the other concepts I have.

A classic case of the dissociation of a concept from its substrate is Parmenides' (who lived in the first half of the fifth century B.C.E.) argument against motion and change. Perhaps the most important remaining fragment of Parmenides' writings is a poem in two parts. In the first part, Parmenides presents a classic argument against the possibility of motion. In this he may -- or may not -- be replying to Heraclitus' argument that everything is in constant flux.[1] His argument is simple: if anything exists, it must be everywhere, and completely homogeneous. In order to move, it would have to go to someplace where there is no thing -- nothing. But nothing doesn't exist.

Parmenides says that there are only things that exist, and whatever doesn't exist doesn't exist. Following from this festival of tautologies, what exists is everywhere, because if there was anywhere where there was nothing, then nothing would exist, but, of course, nothing doesn't exist. Notice that "No thing is there" changes to "Nothing is there." Here the absence of anything has been turned into "no thing" or "nothing," which is a noun, i.e., the name of a person, place or thing. But it is no thing, which is a logical contradiction.

But Parmenides' argument is broader still. He argues that not only motion but change is impossible. In this argument he relies on two principles already widely believed by the Greeks: the principle of non-contradiction and the principle of indestructibility of matter. The first principle says that a thing cannot be and not be in the same sense at the same time, while the second holds that nothing can come from nothing and nothing can pass into nothing. His most famous example is the ripening tomato: first, there is a green tomato, then there is a red tomato. Where did the green tomato go? It can't have passed into nothing -- and where did the red tomato come from? It can't have come from nothing. The conclusion Parmenides reaches is that change is an illusion.

Parmenides' argument had a profound effect on Greek thought. Leucippus (first half of the fifth century B.C.E.), one of the originators of atomic theory, may or may not have been reacting to Parmenides' argument and/or the arguments of Zeno, who argued that motion was impossible because a moving object would have to pass through an infinite number of points, which would take an infinite amount of time. In any event, Leucippus argued that the world was composed of tiny indivisible particles moving through the "void". The notion that the atoms are indivisible means that there is a limit to how finely things can be divided, eliminating the possibility of an infinite division.

Since the atoms are eternal and unchanging, and changes in the appearances of phenomena are solely the result of the arrangement of the atoms, Parmenide contradictions are avoided. As for the existence of the "void" or empty space, Leucippus is said to accept Melissus of Samos' (fifth century B.C.E.) argument that the void is necessary for motion, but, since we observe motion, this proves the existence of the void or empty space.

Leucippus' associate Democritus (ca 460-370 B.C.E.) simply dismisses Parmenides argument against the existence of empty space with what Graham[1] calls an "indifference argument":

a. Thing is no more than not-thing.
b. Thing exists.
c. Hence, not-thing exists

In this argument we can see the growing rift between the two social networks: Parmenides accepts the primacy of reason over experience, and dismisses observational evidence on the basis of his philosophical arguments. Leucippus and Democritus, on the other hand, dismiss the philosophical arguments based on observations.

Parmenides' poem has two parts: the first is the argument just discussed. This section of the poem he refers to as the "Aletheia" or "Truth." In the last part of the poem, however, Parmenides presents his own cosmology, which he means to be the best possible explanation of human experience. This section of the poem he calls "doxa," which is usually translated as "opinion", but which probably means more like knowledge about the world of experience rather than the ideal world of truth. Any theory about actual experience in the world would be called doxa. The separation of the world of experience from the world of thought was common in Greece, and this separation is what makes it possible for Parmenides on the one hand to deny that motion and change was possible, yet on the other hand present a cosmological model meant to explain the motion and change we experience every day. This division is even clearer in the case of Plato.

However much effect Parmenides' argument may or may not have had on the Ionian scientists, its effect on the Athenian philosophers Socrates, Plato and Aristotle was profound.

1.   D. W. Graham, *Explaining the Cosmos: The Ionian Tradition of Scientific Philosophy*. (Princeton University Press, Princeton, 2006).

2.   S. Berryman, in *The Stanford Encyclopedia of Philosophy* (http://plato.stanford.edu/archives/fall2010/entries/leucippus/, 2010.)

Chapter 4: The Athenian Culture

And now, Socrates, Plato and Aristotle begin to construct the philosophy that will serve as the foundation of Christianity and Western Culture. It eschews the principles of Ionian Science in favor of a categorical, perfect, unchanging definition of knowledge -- it is a philosophy that insulates itself from change based on observations.

As we discussed earlier, Greek authors often attributed acts to other individuals whether they did them or not, and Plato is no exception. A good deal of his work he ascribes to Socrates in a series of dialogues with others. In the dialogue *Parmenides*, Plato describes a visit to Athens by Parmenides and his student Zeno. At that time, Socrates was described as a young man, while Parmenides was aging. Plato's high regard for Parmenides is shown by the fact that, in every other dialogue, Socrates wins the argument, but the argument with Parmenides ends in a draw.

Plato (ca 424-327 B.C.E.) completely accepted Parmenides' notion that the appearances of everyday life could not be the source of true knowledge. In the Republic, book VII, Plato makes this emphatic:

> You may be right, and I may be wrong. But I, for my part, cannot think any other study to be one

that makes the soul look upwards except that which is concerned with the real and the invisible, and, if anyone attempts to learn anything that is *perceivable*, I do not care whether he looks upwards with mouth gaping or downwards with mouth closed: he will never, as I hold, learn -- because no object of sense admits of knowledge. (Republic VII, emphasis in original)

Plato presses the Greek distinction between practical knowledge of the world of experience (doxa) and perfect knowledge to its extreme: for Plato, true knowledge, which he calls *episteme*, is perfect, exact, abstract and unchanging. Although Plato continues the tradition of teaching doxa, and produces a cosmology of his own, he believes firmly that nothing perceivable can yield true knowledge (episteme). In fact, the true object of knowledge is not to be found in the world of experience at all, but rather resides in another, perfect world -- the World of Ideas, a mystical land from which we have been expelled for some unknown offense into this world, which is a place of deceit and punishment. What we know of the truth is solely the result of (imperfectly) remembering what we once knew before our fall. Indeed, the word *educate* comes from the Latin *e ducere*, which means "to lead out of," since educating means to draw out what we already knew but is still buried deep in our minds.

Plato's student Aristotle (384-322 B.C.E.) addresses Parmenides' challenge in another way. He agreed that true knowledge or episteme was perfect, universal, unchanging and certain, but he devised a method he believed could gain such knowledge from the world of experience. He conceived of the world as composed of two principles -- *primary matter*, a featureless potential that filled the universe but had no properties of its own, and *substantial form*, which informed matter and made it into actual things. When a tomato ripens, its matter remains unchanged; nothing ceases to be and nothing comes to be from nothing. Rather the matter of the tomato gives up the form *green* and takes on the form *red*.

But where did the forms come from? They come from the *essence* of the tomato, that which makes a tomato what it is, and every form through which the tomato will pass is contained within the essence as potential. What appears as change is the succession of forms moving from *potential* to *actual*, or from *potency* to *act*, as the scholastic philosophers generally translate it.

And before that? The essence of the tomato has its origin in the tomato seed, which itself contains all the potential forms through which the tomato will pass. And before that? The seed, of course, came from an earlier tomato and so on. But for the Athenians, the notion that a dependent chain could depend on nothing was anathema, and so Aristotle concludes that there exists an *uncaused cause,* a being which is at once the cause of all things but is itself uncaused, always existing, and containing within it all the forms through which all things will pass forever.

Aristotle's solution to the problem of motion is similar. For modern readers, long familiar with Newton's concept of empty space, and even more so for scientists familiar with Einstein's curved Riemann space and the strange space of quantum electrodynamics, it is hard to grasp the difficulties of thinking about motion that come about in a culture like Athens where a concept of space has not yet emerged. For Aristotle and the other members of his social network, empty space is inconceivable. Even empirical evidence seems to support this idea for the Athenians. They know that water will not pour from an inverted narrow-necked bottle unless a hole is made in the bottom for air to enter. From this and other evidence they believe that "nature abhors a vacuum."[1]

Aristotle, instead, has a concept of "place," which is the area immediately surrounding an object. Each object has within its essence the potential to be in a number of places. The place it is at the moment is its *actual* place, but it has the potential, i.e., is in potency, to be in several other places. When it moves, it moves from its actual place to the place it is potentially; when it finishes moving it will actually be in the new place. Thus Aristotle defines motion as the act of a being in potency insofar as it is in potency. While moving, it isn't actually anywhere.

As with change, the potency to be in a place must pre-exist the motion; all motion is itself caused by motion. As with the forms, the chain of movers cannot be infinite, and so all motion goes back in the end to the *unmoved mover*, who, of course, turns out to be the uncaused cause.

Because the cause must exist before the effect, motion is always preceded by an intention. Heavy things fall because their "proper place" is the center of the universe, but the intention to be there is in the heavy thing. Fire rises because the proper place for light things is the periphery of the universe. Fire *intends* to rise; the intention must be present for the action to take place. This is known as Aristotle's *entelechy* -- a world in which everything happens for an end.

Aristotle applies his notion of entelechy to human beings as well. People's actions are caused by their *intentions*, which of course pre-exist the action. Not only does Aristotle assert that every human action is motivated by some intention, but that all men have the same goals, which are fixed in their nature. This idea will be the founding premise of Thomas Hobbes' *Leviathan*, which begins by arguing that, since men have the same goals, and are of about the same ability, they are natural enemies:

> From this equality of ability ariseth equality of hope in the attaining of our ends. And therefore if any two men desire the same thing, which nevertheless they cannot both enjoy, they become enemies...[2]

These intentions exist in the mind, which is an immaterial thing, not connected to the body in any way. "...the mind is, according to Aristotle, not "mixed with the body", insofar as it, unlike the perceptual faculty, lacks a bodily organ (*De Anima* iii 4, 429a24–7).[3]

Like physical objects that move, human beings move from one state of being to another state of being. Human behavior consists of discrete jumps from one state to another state. But Aristotle warns against too high a standard for understanding human behavior. In book one, chapter three of the Nicomachean Ethics he begins with a caution:

> Our discussion will be adequate if it has as much clearness as the subject-matter admits of, for precision is not to be sought for alike in all discussions, any more than in all the products of the crafts. Now fine and just actions, which political science investigates, admit of much variety and fluctuation of opinion, so that they may be thought to exist only by convention, and not by nature. And goods also give rise to a similar fluctuation because they bring harm to many people; for before now men have been undone by reason of their wealth, and others by reason of their courage. We must be content, then, in speaking of such subjects and with such premises to indicate the truth roughly and in outline, and in speaking about things which are only for the most part true and with premises of the same kind to reach conclusions that are no better. In the same spirit, therefore, should each type of statement be received; for it is the mark of an educated man to look for precision in each class of things just so far as the nature of the subject admits; it is evidently equally foolish to

accept probable reasoning from a mathematician and to demand from a rhetorician scientific proofs.

It is pertinent that Schrödinger thought the first step toward science, which he found in Thales, was the idea that the subject under study *could be understood*. Here Aristotle warns us in advance that human behavior *can't be understood*. Schrödinger studied the Greeks to search for a concept we inherited from them that was causing a debilitating prejudice preventing progress in theoretical physics. Such a prejudice would be something we believe to be a fact learned from experience when actually it was only a historically inherited prejudice. *Here we see the origin of the widespread prejudice in the social sciences that all human behavior is intentional, and that precise observations and explanations of human behavior are not possible.* While it is commonly believed that these are facts learned through experience, in fact they are historical prejudices invented by Aristotle.

The emphasis on reasoning over observation that characterizes the Athenians is well illustrated by both Plato and Aristotle's concentration on logic. Aristotle's syllogistic is a form of reasoning based on nested categories. In the classic AAA (Barbara) form, a syllogism contains two universal affirmative premises and a universal affirmative conclusion, such as

All dogs are mortal.
All spaniels are dogs.

Therefore, all spaniels are mortal.

When the nesting scheme implied by the syllogism is true (in this example, all dogs are contained within the set *mortal*, and all spaniels are contained within the set *dogs*, it follows that all spaniels must be contained within the set *mortal*, the syllogism is *valid*. If the syllogism is valid and if the premises are true, then the conclusion is true. If the syllogism is not valid, or the premises are not true, then the conclusion is *doubtful*. Aristotle and his medieval successors elaborated the study of syllogisms, identifying 256 types of syllogism.[4]

The main problem with the syllogism, however, is its categorical nature. Syllogistic reasoning is too cumbersome for the analysis of motion and change, and for the description of irregular shapes. Recall that the Ionian scientists' fundamental method of observation is comparison, which is inconsistent with a categorical reasoning scheme.

The resulting Athenian system of observation and reasoning stands in sharp contrast to that of the Ionian scientists. Athenians define knowledge as *episteme*, which means perfect, abstract, certain and unchanging knowledge. In the case of Plato, the object of this knowledge is not this world, but the Ideas in the immaterial World of Ideas. In the case of Aristotle, the source of knowledge is abstract form that is abstracted from the world by induction. Knowledge is categorical, as is reasoning. Reasoning takes priority over observation, and if reasoning contradicts the senses, the senses are not to be believed. All action, including human action, is teleological. Whatever is done is intended. Human behavior particularly is not understandable with the same precision as other aspects of human experience. Everything that happens is set in play by an abstract, eternal being that is the source of all the forms and all the motion in the universe.

These principles represent the foundation of the Athenian's observational and reasoning system. It is a blunt instrument indeed. Without a comparative measurement model, observation of motion and change is crude, and information from the environment is muted. It is a deeply conservative system, insulated against incoming information that could modify the collective neural pattern. In the millennia following, we will see these principles cropping up again as again as "findings" or "laws" in what their authors believe to be original theories, never understanding that the reason they believe these arguments to be true is that, for purely historical reasons, they form the foundation of their cultural beliefs.

Chapter 5: The Athenian Network and Christianity

And now, as the Roman Empire slides into history, Christianity emerges as the dominant cultural force in Europe. Through Doctors of the Church Augustine and Aquinas, it founds its beliefs on the ancient Hebrew texts, writings about Jesus, and the philosophy of Plato and Aristotle. Ionian science is severely repressed.

As important and influential as the Athenian culture was, it took on much greater significance through the work of two later philosophers and theologians, Augustine (354-430 C.E.) and Aquinas (1225-1274 C.E.). In the fourth century C.E., early Christianity was in the process of consolidating its organization and formal doctrines, purging alternative doctrines, heresies and schisms, and purging competing religions such as Judaism, paganism and others. Among the most influential of the early Doctors of the Church (from the Latin *docere*, to teach) was Augustine, the Bishop of Hippo.

Although he converted to Christianity only in his mid thirties, thereafter Augustine produced a prodigious volume of work refuting alternative dogmas and establishing a systematic philosophical basis for Christianity. In this, his most important influences were scripture and Plato. Augustine was strongly influenced by the Neo-Platonism of his time, as were other Christian writers such as Origin. As Mendelson suggests:

One of the decisive developments in the western philosophical tradition was the eventually widespread merging of the Greek philosophical tradition and the Judeo-Christian religious and scriptural traditions. Augustine is one of the main figures through and by whom this merging was accomplished.[1]

Among the most important parts of Athenian culture that Augustine adopted was Aristotle's syllogism:

The rules about syllogisms and definitions and classifications, on the other hand, greatly help people to understand, provided that they avoid the error of thinking that when they have mastered them they have learnt the actual truth about the happy life (De Doctrina, Book Two, XXXVII55-XXXXVIII56).[2]

Augustine also believed that Aristotle's Syllogistic was essential for understanding scriptures:

Dominant are the subjects of logic and number, but logic is of paramount importance in understanding and resolving all kinds of problems in the sacred texts. (De Doctrina Book Two, XXX .47-XXXI.49)[2]

Thus, the tool Augustine chooses to guide his reasoning, not only for his philosophy but for his spiritual thought as well, is the categorical syllogistic of Aristotle.

As important an influence as Aristotle was -- Augustine was trained as a rhetorician and had also read Aristotle's Rhetoric and Poetics -- clearly Plato was the major influence:

> The single most decisive event, however, in Augustine's philosophical development has to be his encounter with those unnamed books of the Platonists in Milan in 384. While there are other important influences, it was his encounter with the Platonism ambient in Ambrose's Milan that provided the major turning point, reorienting his thought along basic themes that would persist until his death forty-six years later.[1]

Augustine's thought is a mixture of the ancient Hebrew books, the New Testament along with other writings about Jesus that weren't included in the New Testament, and the Athenians. It embodies the same principles that define the Athenian way of thinking.

The work of the Athenians had a chilling effect on science, which slowly declined, then disappeared during the first few centuries of the Christian era. As Thomas Kuhn points out,

> Ptolemy, the astronomer, and Galen, the physician, were the last great figures in ancient science, and they both lived in the second century A.D. [C.E.].... When the Moslems (sic) invaded the Mediterranean basin during the seventh century, they found only the documents and the tradition of ancient learning. The activity had largely ceased.... Science was particularly neglected because...the Catholic Church was initially hostile to it.[3]

Indeed, Augustine explicitly advises Christians that they need not worry about understanding the cosmos, geology, biology, meteorology and other early sciences. It is sufficient to believe in God, and know that he created everything that is:

> When, then, the question is asked what we are to believe in regard to religion it is not necessary to probe into the nature of things, as was done by those whom the Greeks call *physici*; nor need we be in alarm lest the Christian should be ignorant of the force and number of the elements, -- the motion, and order, and eclipses of the heavenly bodies; the form of the heavens, the order and the natures of animals, plants, stones, fountains, rivers, mountains; about chronology and distances; the signs of coming storms; and a thousand other things which those philosophers either have found out, or think they have found out...It is enough for the Christian to believe that the only cause of all created things, whether heaven or earthly, whether visible or invisible, is the goodness of the Creator, the one true God; and that nothing exists but Himself that does not derive its existence from Him.[4]

Other contemporary writers are more hostile, not only claiming such knowledge is unnecessary, but attacking it as false and ridiculous.[3] So serious was the decline in secular scholarship that Christendom not only lost the practice of science, it lost the writings -- both Ionian and Athenian -- themselves.

Fortunately, a good bit of the writings ended up in the possession of the Muslims, and after the Muslims invaded the Mediterranean basin in the seventh century C.E., the works began trickling back to the West, first in

Arabic, then slowly translated into Latin, and by the eleventh and twelfth century, interest in the "ancient wisdom" was widespread and intense. Thomas Aquinas (1225 - 1274 C.E.) was the most important of the scholastic writers, and left a huge corpus of philosophical and theological works based primarily on Aristotle (whom he called "The Philosopher") and scripture.

The merger of Plato and Aristotle with Christianity produced a worldview that was highly resistant to change. The Christian world consisted of Aristotle's universe of nested spheres, augmented by Ptolemy and the Muslims to include a ninth sphere which accounted for epicycles. At the center was the spherical earth, and at the outer periphery was the abode of God, which was an amalgam of the perfection of the heavenly spheres and Plato's World of Ideas. God himself was an amalgam of the Hebrew personal God and Aristotle's uncaused cause and unmoved mover, the creator of the universe and source of all motion and change within it. The earth and those things beneath it are, as in Plato's own model, corrupt, with the lowest sphere being Hell, the source of evil in the world.

Within the Christian world stood humans, who were themselves an amalgam of the heavenly (the abstract, non material soul or mind) and the earthly (the body). Of these, the body is evil and flawed, as is everything in the world of experience -- a place of trial and suffering, where we have been exiled due to our "original sin." Life is a perpetual struggle between the evil body, whose lustful appetites drive it toward evil actions, and the divine soul, which must reign in the appetites and focus on spiritual, non-material things. If the body wins, it goes to its proper place -- the center of the universe, or Hell. If the immaterial soul prevails, it goes to its proper place -- the periphery of the universe, the highest sphere, to dwell in the abode of God. In fact, so great is the corruption of the Earth and the humans who dwell on it that the most heroic efforts would fail to win salvation without the assistance of divine revelation, a gift to humans resulting from Jesus' sacrifice.

This model is highly conservative and resistant to change. First, scientific observation is strongly discouraged because the senses are a source of error and confusion, which can prevent God's divine message from being heard and understood. Secondly, the categorical system of observation and reasoning embodied in Aristotle's syllogistic is a blunt tool that excludes comparative measurement, which is the primary method of scientific observation.

As we have seen, the neural pathways in the collective neural network can only be modified in two ways: by direct observations or by symbolic communication. The minimization of direct observations and the course symbolic encoding of such observations resulting from a categorical scheme produce a system that is insensitive to its environment, and determined primarily by symbolic communication -- which, of course, is controlled by the Church.

By the fourteenth century C.E., Aristotle's thought had passed from the special interest of the cloistered academician to the artistic community. Dante's Divine Comedy is a journey through Aristotle's cosmos, as modified by the Muslim astronomers and with the addition of the Christian God, an amalgam of the personal Hebrew God and Aristotle's unmoved mover/uncaused cause. By the sixteenth century, Aristotle's model is the cultural substrate of every European. By now, Aristotle's basic concepts have become the "historical prejudices" sought by Schrödinger -- concepts we believe to be based on solid evidence, but which are really only held because they were erroneously adopted deep in our history.

Shakespeare's heroes struggle because of the appetites that drive them toward ends, sometimes comic, sometimes tragic, but always, underneath, Aristotelian, based on Aristotle's teleological model of human behavior. By the twentieth century, Aristotle's entelechy, though the agency of Shakespeare, gives rise to another psychological model, that of Sigmund Freud. Freud began reading Shakespeare as a very young boy, and reread him several times. His influence over Freud is immense:

> In some sense, Freud has to be seen as a prose version of Shakespeare, the Freudian map of the mind being in fact Shakespearean. There's a lot of resentment about this on Freud's part because I think he recognizes it. What we think of as Freudian psychology is really a Shakespearean invention, and, for the most part, Freud is merely codifying it.[5]

Freud's model of a conscious "mind" struggling to control an "unconscious" populated by suppressed urges pressing for gratification not only mirrors Shakespeare's heroes, but is in fact the Augustinian version of Aristotle's psychology.

By the beginning of the twentieth century, the culture of the Athenian network, merged with the Judeo-Christian culture, had formed the fundamental concept of the nature of human beings and the theory of human thought and behavior that permeated the western world. Westerners believe that human beings are rational animals, impelled by internal appetites to move from one state to another in discrete behaviors, fundamentally unpredictable and not subject to precise scientific understanding. During that century, hundreds -- perhaps thousands -- of "new" theories of human behavior rose up in psychology, sociology, political science, economics, anthropology, communication and elsewhere, all of them fundamentally Aristotle's theory thinly disguised in modern terms like needs, drives, impulses, intentions, attitudes and more, but all, at the root, not the product of careful research, but rather the historical legacy of the Athenian Greeks.

1.   M. Mendelson, in *The Stanford Encyclopedia of Philosophy*, edited by E. N. Zalta (Stanford, 2010).
2.   Augustine, *De Doctrina, Book II.*
3.   T. S. Kuhn, *The Copernican Revolution.* (Harvard University Press, Cambridge, Massachussets and London, England, 1957).
4.   Augustine. (1871-77).
5.   D. Leybman, Interview with Harold Blume in *The Paris Review* (Spring, 1991).

Chapter 6: The Ionian Network of Modern Scientists

And now, although deeply faithful and committed to Christianity, a few still read the ancient Ionian books and make measurements the old, comparative way. So great is the power of the Ionian methods that they compel Copernicus, Kepler, Galileo, and Newton to heretical opinions about the Cosmos; views that survive the wrath of the Church and form the foundation of modern science.

Meanwhile, in sixteenth century Poland in a land ceded by Germany in 1466, the child of a German family struck at the heart of the Athenian culture. On his deathbed, Nicholaus Copernicus held the first copy of his book *De Revolutionibus Orbeum Coelestium*, in which he claimed that the sun, not the earth, lay at the center of the universe, and that the earth revolved about the sun along with the rest of the planets. In this he was following in the footsteps of the Ionians, since Aristarchus of Samos had published a heliocentric model in the third century B.C.E. Copernicus was well aware of Aristarchus' model, since he discussed Aristarchus' model in an earlier draft of *De Revolutionibus*, which survives. He does, however, mention that Philolaus the Pythagorean concluded that the earth moved around the central fire, and that

Heraclitus and Ecphantes (both Pythagoreans) suggested that the earth rotated on its axis.

Copernicus was one of the few scholars that had maintained a "dual citizenship," living in a network dominated by the Athenian culture, but keeping a connection to the Ionians by reading the ancient writings. Although thoroughly familiar with Aristotle's cosmology (Copernicus struggled to maintain as much of Aristotle's model as possible, even in his heliocentric model) he was steeped in the Ionian comparative measurement model and therefore could not ignore the discrepancies between what Aristotle's cosmology predicted and what the corpus of observations revealed. He believed that his heliocentric model was not just a better model, but the "correct" model:

> Copernicus proposed his theory as a true description, not just a theory to save appearances. Unlike Buridan and Oresme, he did not think that any theory that saved appearances was valid, instead believing that there could only be a single true theory. When the work was published, however, Andreas Osiander added an unauthorized preface stating that the contents were merely a device to simplify calculations.... Similar theories had been proposed by Aristarchus as early as the third century B. C., and Nicholas de Cusa, a German scholar, had independently made the same assertion in a book he published in 1440. We know for a fact that Copernicus was well aware of Aristarchus's

priority, since his original draft of *De Revolutionibus* has survived and features a passage referring to Aristarchus which Copernicus crossed out so as not to compromise the originality of his theory. In his belief that his theory was an accurate description of nature rather than just a mathematical model, Copernicus was therefore not truly revolutionary.[1]

Even before Copernicus, scholars at the University of Paris had renewed the Ionian work. Jean Buridan, along with other scholastics, revised and secularized Aristotle's logic. He rejected Aristotle's concept of motion, claiming instead that the force that launches a moving object impresses into it an "impetus" which continues to move it after the original cause has ceased. This impetus will continue to move the object until it has been exhausted by overcoming resistance to movement. (It follows that, facing no resistance, it would continue to move forever.)

Buridan also rejected Aristotle's model of human behavior, picturing a donkey standing equidistant between two equally delicious meals, starving because it could not decide between them -- a tale now known as "Buridan's Ass."

Buridan's younger colleague Nichole Oresme advanced the notion of impetus and helped lay the groundwork for Newton's theory of motion. He rejected Aristotle's concept of place, and laid the groundwork for Newton's concept of an infinite, absolute space (which subsequently was abandoned in favor of Einstein's curved space, but which was necessary for the development of concepts of multiple solar systems and particularly Newton's theory of motion.)

The most famous Ionian of the Scientific Revolution before Newton, Galileo (1564-1642 C.E.) struck at the Athenian model with his relentless use of the comparative method of the Ionians. In his study of buoyancy, he placed a pea in a tank of water and added salt until the pea floated half way between the top and bottom. Then he added a quantity of salt, and recorded that the pea rose by three fingerbreadths. In his study of pendulums, he compared the period of their motion to the beating of his pulse.

Galileo's most famous experiment of the inclined plane rests entirely on the comparative method. The determination of the rate at which objects fall is not possible in a categorical system, particularly if the categories are coarse. An excellent illustration of this is the fact that Aristotle's laws of motion are easily disproved by experiments elementary school children can be taught to perform. Aristotle holds that objects have a "proper speed" at which they move; whenever they move they begin at their proper speed, continue at that speed until the reach their goal, then stop. Heavy objects, he claimed, fall faster than lighter objects, and fall at rates proportional to their weight.

In his book *The Copernican Revolution*, Thomas Kuhn, who had a Ph.D. in physics, says:

Today every schoolboy knows that heavy bodies and light bodies fall together. But the schoolboy is wrong and so is this story. In the everyday world, as Aristotle saw, heavy bodies do fall faster than light ones. That is the primitive perception. Galileo's law is more useful to science than Aristotle's, not because it represents experience more perfectly, but because it goes behind the superficial regularity disclosed by the senses to a more essential, but hidden, aspect of motion. To verify Galileo's law by observation demands special equipment; the unaided senses will not yield or confirm it. Galileo himself got the law not from observation, at least not from new observations, but by a chain of logical arguments...[2]

This is quite wrong. But don't let me convince you with arguments when direct observations are available and easy. Try it yourself. Put a US dime on top of a US quarter, hold the quarter by the edges, and drop them onto the floor. The quarter will not accelerate away from the dime whatsoever, in spite of the fact that it is several times heavier. Stand on your chair and drop them from twice as high, with the same result. Replace the quarter with a US dollar coin. Nothing changes. Replace the dollar with Professor Kuhn's book, with the same result. Replace the book with Webster's unabridged dictionary, which weighs thousands of times more than the dime. Climb up on top of your desk and drop them from the ceiling. They will fall with the

94

same acceleration, and it will be clear to any observer that there is no detectable difference whatsoever. (If you use a really heavy book, when it falls to the floor it will make a very loud noise, and this, plus the fact that you are standing on your desk, makes a demonstration your students won't soon forget.)

Nor did Galileo reason to his law "by a long chain of logical arguments." Galileo performed a series of experiments with inclined planes that are easily reproduced by schoolchildren without special equipment. First, he rolled balls down grooved inclined planes at two different angles of inclination, but the same height. He found that the velocity of the ball once it left the inclined plane was the same, and concluded that terminal velocity depended only on how far it had fallen, and not on the slope.

He next rolled the ball down the inclined plane and measured the distance the ball rolled in equal intervals of time. Some hold that he measured the time with a water clock, a device that continually drips water into a container. The water can subsequently be weighed to determine the elapsed time. Some have suggested that Galileo could not have achieved enough precision with his water clock to derive the law, while others claim he could. Some modern experimenters claim to have used this device and achieved sufficient precision. But, in 1975 -- 18 years after Kuhn's book was published -- Drake suggested that rediscovered pages from Galileo's notebooks showed he used a much more precise method: music.

Galileo's father and younger brother were both professional musicians, and Galileo himself was an accomplished amateur lutenist. Galileo's lutes at that time would have had moveable frets, which were actually strings tied around the instrument's neck. These could be moved back and forth to adjust the instrument's intonation. Drake suggests that Galileo tied strings around his inclined plane, and rolled the ball over them, which produced audible clicks. He could then sing along while the ball was rolling, adjusting the location of the frets until the ball clicked over them on the beat of the song he was singing. Once he had "tuned" his inclined plane, all he needed to do was measure the distances among the frets to determine the law of falling bodies.[3] The level of precision of measure that can be attained by this method is considerably higher than the water clock, and more than sufficiently precise to establish the law.

Most importantly, whatever method Galileo used was inherently comparative. With a categorical measurement system, it is not possible to establish the correct law, which is likely the reason why those who held to Aristotle's categorical scheme could not tell that his laws of motion were false for about 2200 years. Even if people could see that the rate of falling was the same, the Athenian priority placed on reason over observation would have led them to distrust their observations.

Galileo's contemporary, Johannes Kepler (1571-1630 C.E.) was another dual citizen of the Athenian culture and the Ionian network of scientists. Kepler was a deeply religious man with a genuine understanding and respect for the work of Aristotle. He considered himself to be working in the tradition of Aristotle's metaphysics and adding to his astronomy. He deeply believed in the Platonic notion that the heavens were perfect spheres, and that the orbits of the planets must be perfect circles. He spent many years of his life trying to build models of the universe embedded in the five perfect Platonic solids, the tetrahedron, cube, dodecahedron, icosahedron and the sphere, but his efforts resulted in failure.

But, as an astronomer and mathematician, he was also steeped in the comparative measurement model of the Ionian scientists. By comparing the trajectories predicted by his models to the measurements of ancient and contemporary astronomers, he was led to conclude that his models were inadequate. And by comparing the measurements of other astronomers to each other, he was aware that there were problems with the measurements themselves.

Kepler himself was not a good observer because his illness as a child left him without the physical resources needed for rigorous astronomical observations, but he knew that the most accurate observer of all was Tycho Brahe (1546-1601 C.E.). Brahe was a keen observer, but not a theorist. He invited Kepler to stay with him, and, after Brahe's death, he was able, after considerable difficulty, to obtain Brahe's data.

Over a period of many years, Kepler focused his attention on the orbit of Mars. He was able to find a circular orbit that fit Brahe's observations to an average of two degrees of arc (a very small error) but at some points the error was as high as eight minutes of arc. Because of this, and because of the cumbersome nature of his model, he gave up the idea of circular orbits and searched for an alternative. After a great deal of trial and error, he found that the orbit was an ellipse with the sun at one focus, which became Kepler's first law of planetary motion.[4]

Kepler's example shows not only that science depends entirely on the method of comparisons, but it shows that it depends on *precise* comparisons.

From this point forward, scientific progress moves forward very rapidly as the Ionian network grew and prospered. Isaac Newton (1642-1727 C.E.) was born on Christmas day the year that Galileo died. Legend has it that one day he saw an apple fall from a tree and from this deduced his universal law of gravitation. As with most myths, this is not exactly true, as Bronowski makes very clear:

> Let me close by reminding you of what Newton actually did on the day that he conceived $G = kmm'/r^2$. He said to himself "If I throw a ball, it will fall to the ground. If I throw it harder, it will fall a little further off. If I throw it harder still, it will fall still further off. I must be able to throw it just so hard that it falls exactly as fast as the horizon, and then it will go all the way around

the world."..... How long will it take? It is easy to calculate, roughly 90 minutes[3].

Well, of course in 1666 when Newton thought of that, nobody was willing to build expensive pieces of apparatus in order to send men round the world to see whether they came back in ninety minutes. That test was reserved for our highly intelligent generation. Newton did not have any subsidies, grants, funds, Secret Service money. But he had the moon. He said "Of course, I cannot throw a ball round the world, but let me now picture the moon, as if it were a ball which has been flung around the world --- 250,000 miles up, but still, it is up there. How long will it take to go round the world?" Well, now it is more difficult. He knew the value of the earth's gravity at the earth's surface, but he did not know the value of the earth's gravity for the moon. He said "Let us suppose that it is given by an inverse square law. Now, how long will it take the moon to go around?" It comes out at twenty-eight days. As Newton said, "They agree pretty nearly."

Now there is the kind of imaginative conception that we put into the laws of nature.

---

[3] The horizon falls about five feet in 8000 meters (about five miles). Five feet is about the distance an object falls in the first second of free fall, so an object moving 8000 meters per second (about 18000 miles per hour) parallel to the surface of the earth will fall completely around the planet and never hit the ground.

How? When we isolate it from the rest of the universe and say "That is the part of it that is going to count. I am not going to be concerned about the perturbations created by Mars and so on." And of course, Newton's was a tremendous mind. You would never get Newton to say, "It came out right." "They agree pretty nearly," said Newton, not forgetting about Mars and Venus and everything upsetting it all.[5]

The point that Bronowski wants to make in this passage -- one that he makes even more emphatically in his television production *Truth or Certainty* in the *Ascent of Man*[6] series -- is that every observation requires that some aspects of experience be neglected or left out. Even though everything in the universe may be interconnected, in order to study any of it, the rest must be left out, as the gravitational influence of the other planets on the Moon's trajectory is left out of Newton's calculations. For this reason, every observation is always subject to some tolerance, or a zone of uncertainty.

From Kepler's urgent need to reduce the discrepancies between his model and the available observations, we learn that an acceptable theory must match observations *to within the error of those observations*. From Bronowski's description of Newton's inverse squares law we learn that perfect agreement will always elude us.

1. Wolfram, (2012).
2. T. S. Kuhn, *The Copernican Revolution*. (Harvard University Press, Cambridge, Massachussets and London, England, 1957).
3. S. Drake, in *Scientific American* (1975).
4. C. Sagan, in *Cosmos*, edited by C. Sagan (1980).
5. J. Bronowski, *The Origins of Knowledge and Imagination*. (Yale University Press, New Haven and London, 1978).
6. J. Bronowski, in *The Ascent of Man*, edited by J. Bronowski (BBC, London, 2012).

Chapter 7: Ionians Today

And now, Ionian scientists like Planck, Einstein, Feynman and Hawking have left no trace of the Athenians in modern science. Gone are the concepts of absolute truth, permanent, unchanging knowledge and ultimate purpose in nature, replaced by the concept of "modeled reality," or symbolic models that approximate observations to ever increasing tolerances, but which are neither true nor false, but only useful.

Thus did Newton's model of the universe replace the early models of Aristarchus, Aristotle, Ptolemy, Copernicus and Kepler. We do not see a steady, incremental improvement. Although Aristarchus' heliocentric model had the potential for much more rapid progress than Aristotle's model, it was not adopted until Copernicus, 1800 years later. Aristotle's model was modified by Ptolemy, and Ptolemy's model was modified by many known and unknown astronomers until its replacement by Copernicus' model. Kepler's work hugely improved Copernicus' model, and Newton's advances seem so great that we don't consider them a modification of the Copernican model so much as a whole new model.

By now, although Copernicus apparently believed that his model was an actual representation of reality as it existed, others were of the opinion that all models were arbitrary, developed solely to model aspects of our experience that are relevant to our interests. Because of superior printing facilities in Germany, Copernicus' friend Rheticus took the manuscript there to be printed. When he was appointed to a mathematics professorship, he left the work in the hands of another person, Osiander. According to the *Stanford Encyclopedia of Philosophy*,

> Though he saw the project through, Osiander appended an anonymous preface to the work. In it he claimed that Copernicus was offering a hypothesis, not a true account of the working of the heavens: "Since he [the astronomer] cannot in any way attain to the true causes, he will adopt whatever suppositions enable the motions to be computed correctly from the principles of geometry for the future as well as for the past …these hypotheses need not be true nor even probable" (*Revolutions, xvi*). This clearly contradicted the body of the work. Both Rheticus and Giese protested, and Rheticus crossed it out in his copy[1].

Newton, too, clearly believed his system was a true description of reality as it existed. He understood space to exist independently of matter, and conceived of absolute motion, that is, the motion of a body relative to space itself. He devised a "thought experiment" to prove this: partially fill a bucket with water and hang if from a long rope. Wind up the rope and let the bucket go. At first, the bucket will rotate, but the water will not. Soon, however, due to friction against the bucket, the water will begin to rotate. Eventually, the water will be rotating at the same velocity as the bucket, so will appear to be at rest relative to the bucket, but the water will climb the sides of the bucket due to centrifugal force. As the rope reaches the limits of its travel, the bucket will stop rotating, but the water will still climb the sides, proving that it is in absolute motion, and its motion relative to the bucket is irrelevant.

Two hundred years later, Ernst Mach suggested that Newton had only performed half of the thought experiment. The other half is to take the bucket, hang it from the rope, fill it partially with water, then rotate the entire universe around the axis of the rope -- which, of course, would produce the same effect.

By the time of Mach, the process of normal science was chipping away at Newton's model. A vital part of Newton's model was his insistence on the existence of empty space. But Newton was also a major contributor to the notion that light exhibits wave-like characteristics. When white light is shined through a prism, it breaks apart into the colors of the spectrum, and this is taken to mean that the color of light is attributable to its wavelength. When one stone is dropped into water, circular waves spread out from the point of impact, but when two stones are dropped in the water some distance apart, the waves from each bump into each other and create a pattern of interference. Experiments with slits through which light passes show similar patterns of interference. But all the waves we know are transmitted through some medium, like air or water. How is light transmitted through empty space?

The most common mechanism was thought to be the *luminiferous aether*, a substance that pervaded the entire universe, whose only property was that it served as a medium for the transmission of light. In the Athenian model it would be enough to establish the existence of the aether by reasoning alone, but for Ionian scientists it is urgent that the existence of the aether be confirmed by some observation.

Since the earth is in motion relative to the sun, and the sun moves even faster around the center of the galaxy, some reasoned that it would be possible to measure an aether wind or drift. Since the direction of the earth's movement is in constant change, it is possible to rule out the likelihood that it might happen to be moving exactly as the aether might be moving, particularly if tests for an aether wind were made in all directions.

Although the earth moves quite rapidly -- about 30 kilometers per second, this is less than one one-hundredth of one percent of the speed of light, about 300,000 kilometers per second. This means that devices of extraordinary precision would be required. As early as 1877, Albert Michelson began experiments with a device he built, completing his first formal experiments in 1881, but the machine lacked enough precision to detect the effect. Along with Edward Morley, he constructed a far superior machine in 1885, which had sufficient precision to detect the expected drift, but no evidence of the drift was found. The experiment was repeated with superior instruments again and again, and, as late as 1926, D. C. Miller reported that his experimental apparatus atop Mount

McKinley had detected the drift. Miller's results could not be replicated, however, and are considered an anomaly.

By this time, work by many scientists -- H. A. Lorenz, Henri Poincare, Ernst Mach, Max Planck, J. J. Thomson, Joseph Larmor, Albert Einstein and others had made the assumption of a luminiferous aether superfluous, and interstellar space was empty again.[2]

But not for long. Increasingly precise measurements indicated that the speed at which light travels is not only independent of direction, but is independent of the relative velocities of observers. An observer approaching a light source will measure the same velocity of light as another observer receding from a light source and another observer at rest relative to the light source. This problem obsessed a young patent clerk in Bern, Switzerland, to the point of exhaustion. One evening, depressed after a conversation with a friend that failed to yield any solution, as he took the train back to his rooms, he looked back at the large clock on the face of a tower in Bern. As the clock receded into the background, he realized that, if he were travelling at the speed of light, the light from the clock could not catch up with him, and the clock would seem to have stopped, although any clock with him on the train would keep time as usual. From this, Albert Einstein awoke from his depression and realized that the idea of a single, absolute time was impossible.[4]

---

[4] Or not. While this is a wonderful story found here and there on the Internet, it's probably not true, and Einstein himself has said "It is not easy to talk about how I arrived at the theory of relativity. There were so

From the constancy of the velocity of light, Einstein was able to derive equations which could calculate the differences in time for observers in relative motion with regard to each other. They showed that differences would be virtually nil except when the relative velocities approached that of light itself. But the consequences were grave for Newton's model. Since no information can be transmitted faster than light, observers in relative motion cannot communicate among themselves without delay, and will disagree as to the time at which events occur. These disagreements rule out the possibility of establishing *even the order in which events occur.* If one event cannot be unambiguously established to have occurred before, at the same time as, or after another event, clearly the simple notion of causality underlying not only Newton's model but Aristotle's before him cannot be supported. Simultaneity is not an empirical matter, but a matter of definition.[4]

Perceptions of distance are equally variable across moving reference frames by exactly the same equations, and the notion of a homogeneous empty void could no longer be upheld. Newton's universe was completely overturned.

Einstein's accomplishments made him the first celebrity scientist of the modern era, and I have no intention of diminishing his achievement. But the

---

many hidden complexities to motivate my thought." 3.
W. Isaacson, *Einstein: His Life and Universe.* (Simon and Schuster, New York, 2007).

problems created by the constancy of the speed of light were of such critical importance to scientists of the era that no stone would have been left unturned in a collective effort to come to grips with them. After the Michelson-Morley experiments failed to detect the luminiferous aether, several scientists sought explanations by investigating the form that observations -- particularly of the shape of Maxwell's electric and magnetic fields -- might take for observers moving relative to one another. Work by Voigt, Heaviside, Thomson, Searle, Lorentz, Larmor, Poincare and others had already established the transformation (now usually called the Lorentz transformation or the Lorentz-Fitzgerald transformations) across reference frames that Einstein derived from his special theory of relativity.

Einstein's discovery came at a time when he was working as a clerk in the Swiss patent office. This was the historical period when steam engines led to railroads which led to the need to keep a general time across long distances. Prior to this period, different localities established their own local times. But if one were to travel from Paris to Berlin by train, one would have to know not only the time in Paris when the train left, but also what time it would be in Berlin.

Today, world time is kept in sync by a network of orbiting atomic clocks in constant communication with each other and clocks on the surface of the earth. But in the beginning of the 20th century, making two clocks only a few blocks apart keep the same time was very difficult. Switzerland was at the forefront of timekeeping, and Bern maintained a system of clocks whose time was coordinated to a central master clock by underground pneumatic tubes, which would send out a pulse of air periodically to synchronize the clocks. Of course, the time lag of the air travelling underground would mean the slave clocks would always be behind the master clock by differing amounts, since they were different distances from the master. A disagreement of one minute would be very good at the time. Consider the difficulties of coordinating clocks across an entire country, or continent.

Among the most numerous claims submitted to the Swiss patent office were systems for synchronizing clocks. The young Einstein would have read many of these, and the notion of synchronizing distant clocks would have been at the forefront of his mind.[5] This happy circumstance would have given Einstein a key advantage in formulating the Special Theory in the way he did, but a thorough review of the literature preceding 1905 shows that the concept already existed in the collective neural network of scientists of the time.

Just as the world was struggling to develop a practical system of timekeeping, it was also coming to grips with the promise of electricity. Since Humphry Davy had first passed an electric current through a strip of platinum and produced a dim glow in 1802, scientists and engineers had been seeking ways to make brighter, longer lasting light bulbs with less electrical energy. In 1894, Max Planck, a professor of physics at the University of Berlin, was hired as a consultant by an electric company to improve the efficiency of light bulbs. Planck was particularly interested in thermodynamics, which is the study of heat and its relationship to other forms of energy. He knew that all matter emits radiation, and that a body that absorbed radiation of all frequencies uniformly -- a "black body" --would itself radiate energy with a frequency that varied with its temperature. There is no perfect black body in nature, but some substances, like graphite, approximate black bodies. A box built of graphite with a small hole in one side will emit radiation when heated that approximates black body radiation. Planck hoped to discover how the intensity of the radiation given off by the body was related to the temperature of the body and the frequency of the radiation.

Planck was a conservative person who adhered to Newton's model of the universe. He was averse to modern ideas, such as Ludwig Boltzmann's belief that heat and energy were statistical phenomena subject to probabilities rather than strictly causal laws. But his efforts to find the functional relationship among intensity, frequency and temperature that satisfied extensive observational data met with repeated failures. Finally, in desperation, he developed an equation that fit the observations to within measurement error, but his equation contained two concepts he abhorred: first, it included a constant from Boltzmann's statistical model, and second, it implied that energy was not a continuous concept, but appeared to be emitted only in discrete packages, each of which was later called a "quantum." He called this "an act of despair ... I was ready to sacrifice any of my previous convictions about physics."[6] Plank was awarded the Nobel Prize in physics for this act of despair in 1918, and Planck's constant h is often considered the foundation of quantum physics.

Continued research in the area of quanta led to an increasingly strange, but highly successful, model of experience. Five years after Planck's desperate publication, Einstein published a paper in which he suggested that light not be considered a continuous expanding wave front, but rather a burst or quantum of energy[7]. These quanta were later called "photons" by Gilbert N. Lewis in 1926.

These quanta behave in ways for which there is no counterpart in everyday experience. In the everyday world in which we live, suppose we position a device that emits small objects -- say, raisins. Some distance away we position a surface which will be marked as the raisins strike it. Between the two, we erect a wall with a window in it. A certain number of raisins will go through the window and strike the surface. If we make a second window in the wall, more raisins will go through and make more marks on the wall. But this is not the case in the world of the truly small.

If a device that emits photons (or electrons, for that matter) is placed a given distance away from a photon detector and a screen with a small hole is placed between them, a given quantity of photons will be detected and counted by the detector. If a second hole is opened in the screen, either more *or fewer* photons will be detected, depending on the distance between the holes. Thousands of other examples of strange quantum behavior can be made.

No physical mechanism has been devised that can account for this strange behavior, and as Richard Feynman says, physicists have given up trying to find one:

> I have pointed out these things because the more you see how strangely Mother Nature behaves, the harder it is to make a model that explains how even the simplest phenomena work. So theoretical physicists have given up on that.[8]

Instead, the path that each photon might take from the source to the detector is modeled as a complex number or vector. The magnitude of the vector is the square root of the probability of the event occurring. The angle of the vector is given by the frequency of vibration of the particle multiplied by the time it takes to traverse the path to the detector. In the case that only one step is involved in the journey of the photon from source to detector, all the vectors representing all the possible paths the photon might take are added together, and the square of the length of the resulting vector is the probability that the photon will be detected. If more than one step is involved -- if, for example, the photon must cross through two surfaces -- then the vectors are multiplied. The square of the length of the resulting vector is the probability of the event. Even though there is no physical model, this mathematical model -- Quantum Electrodynamics or QED -- provides the most accurate correspondence with observations of any scientific theory ever devised.[8]

This progress, however, represents a complete abandonment of every shred of Aristotelian thinking. Gone is the entelechy -- the idea that everything happens for a purpose. Gone is the notion of perfect knowledge through causes. Ionian scientists understand that scientific knowledge will never be perfect, and even causality itself is a victim of relativity theory and quantum physics. So, too, is the concept of "understanding" very different from Aristotle's: what we have instead are mathematical models that fit the results of experiment. As Hawking and Mlodinow argue in their 2010 book:

> "...it is pointless to ask whether a model is real, only whether it agrees with observation. If there are two models that both agree with observation, like the goldfish's picture (of the world, through the distorted lens of a spherical fishbowl) and ours (without this lens distortion), then one cannot say that one is more real than the other. One can use whichever model is more convenient in the situation under consideration. ... Model dependent realism applies not only to scientific models but also to the conscious and subconscious mental models we all create in order to interpret and understand the everyday world. There is no way to remove the observer – us– from our perception of the world, which is created though our sensory processing and through the way we think and reason. Our perception – and hence the observations upon

which our theories are based – is not direct, but rather is shaped by a kind of lens, the interpretive structure of our human brains."[9] (p. 46)

Plato would be spinning in his grave.

1.   S. Rabin, in *Stanford Encyclopedia of Philosophy*, edited by E. N. Zalta (Stanford, 2010).
2.   L. S. Swenson Jr., in *Encyclopedia of Physics*, edited by R. J. Lerner, Trigg, George L. (VCH Publishers, Inc., New York, Weinheim, Cambridge, Basel, 1990).
3.   W. Isaacson, *Einstein: His Life and Universe*. (Simon and Schuster, New York, 2007).
4.   M. Jammer, *Concepts of Simultaneity from Antiquity to Einstein and Beyond*. (The Johns Hopkins University Press, Baltimore, 2006).
5.   P. Galison, *Einstein's Clocks, Poincare's Maps*. (W. W. Norton & Company, 2003).
6.   Anonymous, Max Planck in *Wikipedia* (2012).
7.   A. Einstein, Annals of Physics **17** (1905).
8.   R. P. Feynman, *QED: The Strange Theory of Light and Matter*. (Princeton University Press, Princeton, New Jersey, 1985).
9.   S. Hawking and L. Mlodinow, *The Grand Design*. (Bantam Books, New York, 2010).

Chapter 8: Revolution?

And now, developments in science occur so rapidly we call them the Scientific Revolution. But underlying the advances lie the ancient roots of modern scientific concepts which follow from the Ionian comparative method: distance, time, velocity, acceleration, mass, force, power, energy - - space.

The history of science certainly does not show steady improvement. Aristotle presents a highly articulated theory of motion and change, space, time, the structure of the universe and the place of human beings within it. But it is clearly a divergence from the path of science. With the collapse of the Roman Empire and the rise of Christianity, the Athenian network flourished while Ionian science was suppressed. Certainly sociological factors have a substantial effect on the progress of science.

And the changes from a heliocentric cosmology to the central fire of Philolaus, then to the heliocentric model of Aristarchus and Copernicus represented very large changes, particularly if we take into account the cultural and religious consequences of decentralizing humanity and making us the inhabitants of just another planet. Certainly the change from a finite little cosmos made of eight or nine spheres, each in direct contact with its neighbor(s), to the infinite expanse of empty space with many stars and solar systems of Newton was a large change. And our most recent shift away from infinite, isotropic, flat, empty space of Newton to the curved strange space of general relativity and the perhaps even stranger space of the quanta, where objects don't move along a path from one position to another, but rather move(?) a little along all possible paths as a probability distribution, would have to recognized as an immense change indeed.

Whether to call these changes "revolutions" as Kuhn[1] does is perhaps a matter of personal style, since the word "revolution" itself has no technical meaning. After all, you can find a revolution in hair care at your corner drugstore. A better term might be "evolution," particularly in that we see two competing cultures -- the Ionian and Athenian -- struggling for survival through changing circumstances. But, through all these changes, large as they may be, a line of elements can be seen to endure while developing to higher levels. These include both the core concepts of science, and its fundamental methods. These are not unrelated.

Among the concepts that endure as fundamental to our model of experience are matter, shape, distance, time, motion and change. These concepts are only crudely defined in the seventh century B.C.E., with Thales recognizing only the four substances fire, air, water and earth. These change into one another, but the process is undefined. Later Anaximines improves the concept of change with his concept of compression and rarefaction. Generally all the Greeks recognized the sphere as a fundamental shape, and held that the sun, moon, planets, the heavens and the earth itself were spheres. Aristotle only recognized two kinds of motion, straight and circular, and had no concept of space at all, but only recognized "place" as the area immediately surrounding a body -- where there was no body, there was nothing.

The development of the concept of coordinates made it possible to conceive of an infinite space in which every point could have a unique name which could be discussed and communicated about unambiguously. Thanks to the Hindu-Arabic numbering system, the process for generating the names is simple and systematic, so it is not necessary to remember the names. Any arbitrary shape, either static or moving, can be described and communicated by naming the points that describe it. Changes in the shape are described by functions of the points with time, which can also be represented by coordinates. With the development of alternative geometries and tensor analysis, the space can be arbitrarily warped, yet transformations from any curved space to any other curved -- or flat-- space can be described precisely.

Secondary variables, such as acceleration, force, mass, energy, work and power waited much longer to be added to these early concepts, and the Greeks had no notion that the energy required to spin the orb of the heavenly spheres around the earth would vastly exceed that required to rotate the earth relative to the heavenly sphere -- particularly since the concept of energy had not yet emerged. Even the most sophisticated of modern concepts, such as the law of least action or quantum path integrals, gauge theories and symmetries are developments of these primitive concepts that have endured across all the "revolutions."[5] The result is a finely articulated conceptual system which can describe events of tremendous complexity, and which enables scientists to communicate about them with great precision.

The array of events and processes that can be described and communicated in this Ionian comparative symbolic construction vastly exceeds the impoverished categorical world of the Athenians, which can only say what an entity was and what it became, but can say nothing about the transformation itself; it can say where an object was and where it ended up, but can't describe the ·movement. That the highly developed Ionian concept of space of modern science can deal with discrete things is made evident by the fact that Quantum Electrodynamic theory (QED), which deals with discrete particles, may be the most successful theory in all of science, but the converse is not true: the discrete concept of "place" of the Athenians is helpless in the face of even the simplest motions and changes.

These concepts were developed to describe and communicate about things that move and change. Conception of movement and change without them is extremely crude. By and large, we'll see that the

[5] The story of the historical development of all the important concepts of science is told by Max Jammer in his extraordinary series of books *Concepts of Space, Concepts of Mass, Concepts of Force, Concepts of Time, Concepts of Simultaneity* and others.

Athenian network of social scientists does not have these concepts available to them, and their characterization of human cognition and behavior is consequently extremely crude.

There are constants of methods which endure and develop. First among these is the notion that all observations consist of comparisons of one aspect of experience to another. The divergence of the Athenians from this Ionian principle gave objects an absolute significance, but eliminated the possibility of precise observation. Indeed, the principle *concepts* of Ionian science are comparative, not categorical.

Another key development in the Ionian model is the disappearance of both intentionality and causality. To be sure, the entelechy -- the notion that every action is purposeful, that is, done for an end -- is an Aristotelian concept in the first place, as is the notion that perfect knowledge consists of knowing the causes of an object or event. But, by the middle of the twentieth century, the idea that nature acts purposefully is clearly gone, and the collapse of the absolute nature of simultaneity made a causal interpretation of events impossible. The adoption of statistical thermodynamics and the probabilistic interpretation of quantum phenomena eliminated any simple notion of causality from that arena.

We also see the firm establishment of the notion of *modeled reality* growing through the advancement of science. We saw the dispute over whether Copernicus' model was real or a convenient fiction played out in his most important book. Kepler spent a considerable portion of his life trying to build actual physical models of the solar system out of the perfect platonic figures without success. By the twentieth century, even the concept of physical models was replaced by mathematical models. Maxwell's formal mathematical model of Faraday's field theory model was a giant step in the evolution, as was the notion of modeling quantum events as complex numbers. In this we can see science moving away from the Pythagorean idea that the world is made out of numbers to the modern conception that experience can be represented by numbers.

Another continuing development is the demotion of humanity from the crown of creation to an ordinary part of nature. In Thales' cosmology, as in Aristotle's, humanity lay at the center of the universe. By the fifth century B.C.E., Philolaus had moved us out of the center of the universe, and, by the twentieth century, our concept of the universe had increased so vastly that Shlovsky and Sagan reasoned that, since the only example of intelligent life we know -- humanity -- resides on a minor planet in the far regions of the spiral arm of a minor galaxy, we ought to consider that particular life form mediocre.[2]

These processes may well be the result of accidental interplay of many factors -- temperature increases in Europe from extensive burning of wood, sudden population growth leading to migration from rural to urban areas, increased population density in coastal areas, increased sea trade, development of manufacturing, and many other factors. But the central factor shaping the development of modern science can be seen to be the competition between the Athenian and Ionian concept of the world and our place in it. As circumstances shift, leading first one then the other to increase in prominence, the concepts and methods underlying the one and the other rise and fall.

Built into the Ionian model of comparative measurement are the germs of space and time, and their derivative concepts, force, mass, energy, work, power and the like. These provide a flexible model that can be molded to fit infinitesimally small changes and irregular shapes. Comparative concepts can be formed into models that can account for subtle differences and changes. They make possible the study of motion and change. In contrast, categorical models inevitably yield hierarchies of nested categories. They are well adapted to perfect, unchanging concepts. They work with triangles and squares and spheres and even dodecahedrons and icosahedrons, but fail miserably when faced with irregular or changing shapes. They provide a solid philosophical foundation for absolute hierarchical organizations like the Catholic Church.

1.   T. S. Kuhn, *The Structure of Scientific Revolutions.* (University of Chicago Press, Chicago, 1962).
2.   I. S. Shlovski and C. Sagan, *Intelligent Life in the Universe.* (Doubleday, New York, 1980).

Chapter 9: The Final Frontier

And now, as Ionian science pushes back the boundaries of ignorance and superstition that surround our understanding of the physical world, it struggles to gain a foothold in the study of humanity itself. But the dead hand of Plato is not to be taken lightly, and progress is slow...

Although I have confined the discussion of the advancement of Ionian science only to physics and cosmology, rapid progress took place along a broad front of disciplines, too large to include in a single book, including great advances in chemistry, biology, genetics, computer science, geography, meteorology and many other of what we now generally call the physical and biological sciences. Still, science was slow to turn to the study of human cognitive and cultural processes. Scholars like Robert Grosseteste (1168-1253), Roger Bacon (1214-1308) and Duns Scotus (1284-1350) wrote comments on Aristotle's *De Anima*, his discourse on the human mind, but made no observations. Scholars like Jean Buridan (ca. 1300-1358) shied away from the study of the mind, which he believed to be immaterial and hence not the domain of the natural philosopher.

Enlightenment scholars like Thomas Hobbes (1588-1679), John Locke (1632-1704), David Hume (1711-1776), Jean Jacque Rousseau (1712-1778), Adam Smith (1723-1790)), whose Aristotelian belief that all human behavior is directed by self interest is the foundational assumption of economic theory, and Immanuel Kant (1724-1804), wrote about human cognitive processes, but, while moving generally toward a more secular view of human thought and behavior, conducted no systematic observations. While they contributed little to the development of a science of human cognitive and cultural processes -- Kant, in fact, argued strongly that the empirical study of mental processes was impossible -- they did serve a useful purpose in weakening the grip of Aristotle on Western academic culture. Hume's rejection of Aristotle's notion of the efficient cause was a precursor of later work in relativity theory and quantum mechanics which further undermined the simple Aristotelian understanding of causality, but his good friend Adam Smith's belief in self interest as the basis of human economic activity thoroughly ensconced Aristotelian philosophy at the foundation of Economics.

In general, the social theories of the Enlightenment were based on Aristotelian notions about the nature of Man, and deductive reasoning from those premisses. They commonly assumed that man had built into his nature certain drives and needs which impelled him to action (remember, women were not yet considered rational). Reasoning from these premisses, Enlightenment scholars like Hobbes, Rousseau and Hume discussed what man must have been like in the "state of nature" before society. An early critic of these views, and precursor of Durkheim and Mead in this, was Adam Ferguson (1723-1816) who succeeded David

Hume as Librarian of the Advocate's Library in Edinburgh:

> As against our unscientific conjectures about how we would have felt in a society profoundly unlike the only one we have ever lived in, Ferguson commends the use of historical records. He talks disparagingly about boundless regions of ignorance in our conjectures about other societies, and among those he has in mind who speak ignorantly about earlier conditions of humanity are Hobbes, Rousseau and Hume in their discussions of the state of nature and the origins of society.[1]

By the end of the eighteenth century, the explosive achievements of science, and with them massive industrialization and urbanization led some to advocate the expansion of science to the study of human cognitive and cultural processes.

In Germany, a new discipline that would be called psychology was also emerging. Ernst Heinrich Weber (1795-1878) was trained in medicine, and investigated the tactile senses, particularly perceptions of weight and mass. He used a comparative method for measuring the perception of weights, and found that the just noticeable difference (jnd) was proportional to the percentage difference rather than the absolute weight. Gustav Fechner (1801-1887) was professor of physics at the University at Leipzig, and carried on Weber's work. He called Weber's discovery that the just noticeable difference of weights was proportional to the percentage difference rather than the absolute weight *Weber's Law*. He went on to show that the perceived difference in stimuli was related to the logarithms of their differences, which has generally been referred to as the Weber-Fechner Law, or just Fechner's Law.

The new science of psychology was moved strongly forward by Herman Helmholtz (1821-1894). Trained in medicine primarily as a physiologist, Helmholtz served initially at several universities in what is now Germany as a professor of physiology and anatomy, but concluded his career at the University of Berlin as a physicist. Prior to Helmholtz, some thought that the nerves were malleable, like wax, and stimuli from the senses actually molded their image onto the nervous system. Helmholtz suggested rather that the brain constructed symbols which represented the stimuli, but need have no actual physical resemblance to the stimuli. These "signs" are inferences from past experiences as to how various stimuli "go together". Any single stimulus can only be interpreted in relation to the rest of the stimuli. So, for example, if I pinch a single grain of sand between my thumb and forefinger, both my thumb and forefinger will report a stimulus. If I lay two grains of sand on the surface of my table and press one with my thumb and another with my forefinger in exactly the same position, both finger and thumb will again report the same stimuli. Yet I understand the first pair of stimuli to represent only one grain of sand, and the second pair to represent two grains of sand.

From this, Helmholtz reasons that the conception we have of the world is a symbolic construction and not merely a passive encoding of stimuli. Our notion of space itself is a construction of this sort. In 1868, Helmholtz delivered a lecture titled "On the Actual Foundations of Geometry," in which he takes up the question of non-Euclidean geometries, following Bernhard Riemann.

Riemann (1826-1886) was a student of Karl Friedrich Gauss, who mathematicians generally agree was the first to understand the possibility of non-Euclidean geometry -- that is, geometry that did not make use of Euclid's postulate that parallel lines never meet -- although the first published paper concerning non-Euclidean geometry was that of Janos Bolyai in 1832. The result of eliminating the parallel lines postulate opens up the possibility of geometries of curved space. Riemann's inaugural lecture "On the Hypotheses Underlying Geometry" set the foundations of Riemannian geometry, and Helmholtz wanted to add onto Riemann's work. This is emphasized in a subsequent paper "On the Facts Underlying Geometry," which was deliberately parallel to Riemann's title, but substituted the word "facts" for the word "hypotheses" to emphasize Helmholtz's belief that our notions of geometry rested on empirical observations, and was learned, rather than being *a priori* in the nature of the human mind.[2]

Ernst Mach (1838-1916) was an Austrian physicist and philosopher who was strongly influenced by Fechner and Helmholtz, and believed that the greatest advances in understanding physics would come from a better understanding of human sensory mechanisms. He extended Helmholtz's notion of learning from the individual lifespan to the Darwinian concept of natural selection, and argued that, through natural selection, the sensory mechanisms developed what would seem to be a priori capabilities, although they were a posteriori at the time they developed. Thus, what the senses perceive is not simply a passive image of reality, but rather the senses do considerable "pre-processing" before sending the result to the brain.[3]

We do not see the world as it is, according to Mach, since that would be chaos. Through a process of natural selection, we have evolved sensory mechanisms that select those features from the environment that maximize our chance of survival. This requires an organization of sensory data into a system. Stimuli themselves have no meaning, but are only meaningful in their relationships with all other sensations. This organization of sensory data is not found in the observations themselves, but is the result of thousands of generations of natural selection. Although the mechanisms underlying the process by which sensory data are organized into a network are only now being described, Mach's work is considered a precursor of neural networks.[3]

Mach's highly nuanced view of the relationship between observations and the mental models based on those observations defies simple categorization. Many writers simply call him a positivist, but his views share virtually nothing with the positivism of August Comte, and even less with contemporary social science definitions of the term. Einstein considers Mach to be his precursor, primarily because of his recognition that only relative motions can be established by observation.

Mach's influence over the philosophy of science was immense, and some think of him as the founder of philosophy of science.[3] Certainly he was the focal point around which early twentieth century philosophers of science revolved. Philipp Frank, Hans Hahn, both students of Ludwig Boltzmann, and Otto Neurath, a sociologist from Vienna, began meeting to discuss philosophy and science as early as 1908. They were committed to spreading the work of Mach more widely, particularly his efforts to prevent the inappropriate use of metaphysics in science. They also intended to bring the work of Poincare and others into the conversation.[4]

Discussions were rudely interrupted by World War I, but resumed in 1921. Shortly thereafter, Moritz Schlick, a student of Max Planck and at that time Professor and Chair of the Inductive Sciences at Vienna, joined the group. The group formally incorporated itself as the *Verein Ernst Mach* or *Ernst Mach Society* in 1928 with Schlick as its president. The group was dedicated to the dissemination of scientific ways of thinking, and a year later evolved into the *Vienna Circle*. Among the topics most discussed was the newly published Tractatus by Ludwig Wittgenstein. Seven years in the writing, and years longer finding a publisher, the Tractatus basically dismisses philosophy as simply a critique of language, a message that fit well with the inclinations of the group.

From the outset there were divergences in the Circle, with some -- notably Hahn, Neurath and Rudolph Carnap -- opting for aggressive attacks against Metaphysics on all fronts, and others, such as Schlick, preferring a more academic program of research. Although there was considerable diversity of opinion in the Circle, their body of thought was usually called *Logical Positivism*, although some of the members, like Carl Hempel, rejected the term positivism because of its association with August Comte's earlier, much different and less informed, use of the term, and preferred instead *Logical Empiricism*.[5]

Logical Positivism was primarily concerned with verification and verifiability. A central concern of positivism is that scientific knowledge be connected to observations. An acceptable scientific theory must make predictions that can be confirmed or rejected by observations. A major criticism of metaphysics is that there is no connection between its main terms and any observations that might be made. In its strongest form, Logical Positivism holds that any statement which contains no empirical referent is meaningless, thus defining the whole of the metaphysical enterprise as meaningless (mathematics and logic, which contain no empirical referents, are considered tautologies, and thus immune to the meaningless criterion).

The common wisdom is that Logical Positivism came to a bad end because the statement "any statement which contains no empirical referent is meaningless" contains no empirical referent and is hence meaningless. This is too facile, however. First, there are many streams of work within the broad awning of the Vienna Circle, and these can stand on their own merits. Secondly, the group was dispersed by political forces due to the rise of fascism in Europe more than any other single factor. Of course, they did fail to prove that metaphysics was meaningless, at least in a way that was convincing to metaphysicians, who still abound.

In addition to the members, there were many more loosely associated with the Circle. Two prominent sociologists who were loosely affiliated observers were Paul Lazarsfeld and Marie Jahoda, both of whom played a defining role in the development of the methodology of sociology through their research, teaching and their widely influential textbooks. Thomas Kuhn's book, the Structure of Scientific Revolutions, was published as a part of the official corpus of Vienna Circle works, and has been widely influential in the social sciences.[6]

Another observer of the group who was never a member was Karl Popper, who later became a critic, and whose work would be fundamental in shaping the methodology of the social sciences. Popper was attracted to Marxism and Freud early in his career, but realized later that both Marxism and Freudianism shared a common trait: they could explain any occurrence whatsoever, even contradictory occurrences. Sidney Hook, a philosopher at New York University, organized a conference of psychoanalysts and philosophers in New York City in 1958 to discuss Freud's theory and psychoanalysis. Among his concerns was that one of Freud's central theses, the notion of the

"Oedipus Complex", made equivocal predictions about subsequent outcomes.

The concept of the Oedipus Complex is based on the idea that male children see their fathers as rivals for the love of their mother. According to Freud, this could lead to three possible outcomes: first, because the father is a rival for the love of the mother, the child could grow up to hate his father. A second possibility, however, exists: because the son hates his father, which is a horrible and unacceptable state of affairs, the child might experience a "reaction formation" and overcompensate by an exaggerated love of his father. But wait, there's more...Yet a third possibility arises. Because the notion of hating one's father and loving the rival for the love of one's mother are both unacceptable, the boy might instead withdraw from the situation and become indifferent to his father.

Hook accosted psychoanalysts at the conference and asked them this question: what would a man who did not go through the oedipal stage look like? Most psychoanalysts could not reply, other than to say that would be impossible -- every man must pass through the oedipal stage. Some were insulted by the question, saying they had flown thousands of miles only to be harassed and embarrassed by Hook. Only one provided a substantive answer, which Hook found unsatisfactory.[7]

Both Hook and Popper came to the conclusion that Freud's theory was not a scientific theory, because it could not propose any situation in which observations could falsify it. *Whatever happened* could be explained post hoc by the theory, even contradictory outcomes.[6] Popper, who was early in life an enthusiastic Marxist, later came to the same conclusion about Marxism: the theory did not make any unambiguous predictions, and therefore was flexible enough to account for anything that might happen, and so was not falsifiable.

Popper's conclusion was that only theories that were capable of falsification counted as scientific theories, which is, after all, a rewording of Logical Positivism that escapes the main critique: it does not define statements without unambiguous empirical referents as meaningless, but simply defines them as unscientific.

In the end, Popper asserted that the truth of scientific (or any other) proposition can never be verified. No matter how many times a theory has been verified, there is no guarantee that it will not be shown wrong in the future. Popper spent considerable effort investigating the possibility that repeated confirmations of a theory might lead it to be "closer" to true, and developed the concept of "truthlikeness" or *verisimilitude* to describe a theory that has a strong track record of confirmations, but later accepted as

---

[6] William H. Sewell, no friend of Freud, provided important empirical evidence against Freud's theory. He showed that there was no significant difference in personality among adults whose early childhood socialization was strict or lenient. 8. W. H. Sewell, P. H. Mussen and C. W. Harris, American Sociological Review **20** (April) (1955).

valid criticisms that had been leveled against this concept, which he abandoned.[9]

In the end, he held that no theory could ever be found to be true by induction (or any other method, for that matter), but conceived of a process by which theories could be replaced by "better" theories through natural selection. Although it may be impossible to show theories to be true, it is possible to show them false. Through a process by which theories are rigorously tested, less "fit" theories might be eliminated, while those that survive the relentless, rigorous program of testing will be more fit and thus "better" theories.[10]

Popper's procedure is a precursor of the contemporary idea of "modeled realism", but the notion of modeled realism does not rely on concepts of truth or falsity -- it has no expectation that any model is "true" in the classic sense. They are all just models, some better than others, but, like all models, are neither true nor false.

Moreover, Popper's theory remains categorical, in that theories either meet the test or fail. Modern scientific practice understands that theories are not simply wrong, but wrong by some amount, and the error defines a "residual term" that must be minimized. Thus a failed theory doesn't just say "oops" and go away, but rather tells us something about how the theory needs to be modified to become more accurate. It's hard to overstate the importance of this idea. When Kepler rejected the notion of circular orbits, he did so because he could calculate the amount and direction of deviation from the predicted orbits, and he could show that elliptical orbits reduced those deviations. When Planck's calculations for black body radiation were wrong, they were wrong by a specific amount and in a specific direction. He was able to understand how the theoretical equation had to be modified to reduce these deviations, even though the consequences of such adjustments were unpalatable to him. When observations of interacting electromagnetic fields differed from Newton's predictions, they did so by a specific amount and in a specific direction. Lorentz and Fitzgerald were able to write the precise transformation needed to make the predictions conform to observations even before Einstein had articulated the theory underlying them. Without the notion of *wrong by a measurable amount*, the scientist is left without a clue as to how to choose an alternative theory from the infinity of possible choices.

1.   A. Broadie, in *The Stanford Encyclopedia of Philosophy*, edited by E. N. Zalta (Stanford Stanford, 2009).
2.   L. Patton, in *The Stanford Encyclopedia of Philosophy*, edited by E. N. Zalta (Stanford, Stanford, 2008).
3.   P. Pojman, in *The Stanford Encyclopedia of Philosophy*, edited by E. N. Zalta (Stanford, Stanford, 2011).

4.   T. Uebel, in *Logical Empiricism -- Historical and Contemporary Perpectives*, edited by P. Parinni, W. Salmon and M. H. Salmon (University of Pittsburgh Press, Pittsburgh, 2003).

5.   J. Fetzer, in *The Stanford Encyclopedia of Philosophy*, edited by E. N. Zalta (Stanfore University Press, Stanford, 2012).

6.   T. Uebel, in *The Stanford Encyclopedia of Philosophy*, edited by E. N. Zalta (Stanford, Stanford, 2012).

7.   S. Hook, *Psycholanalysis, Scientific Method, and Philosophy.* (New York University Press, 1959).

8.   W. H. Sewell, P. H. Mussen and C. W. Harris, American Sociological Review **20** (April) (1955).

9.   S. Thornton, in *The Stanford Encyclopedia of Philosophy*, edited by E. N. Zalta (Stanfore University Press, Stanford, 2011).

10.   K. Popper, *Logic der Forschung.* (Julius Springer, Vienna, 1935).

Chapter 10: Psychology

And now, as the social sciences proliferate a wide variety of non-scientific and pseudo scientific theories, a thread from Helmholtz to Mach to Wundt to Durkheim to Mead carried the Ionian approach to social science into the 20th Century.

There was a second branch that sprang from the work of Weber, Fechner and Helmholtz. Helmholtz's assistant at Heidelberg, Wilhelm Wundt (1832-1920), who was the first person to call himself a "psychologist", founded the first psychology laboratory at the University of Leipzig, edited the first psychology journal, and wrote the first psychology textbook. Wundt took his degree in medicine at Heidelberg before becoming assistant to Helmholtz there. He was inspired by the work of Weber and Fechner, and, like his predecessor Jean Buridan 500 years earlier, felt that the human soul was irrelevant to science, since science only dealt with material things. Weber's and Fechner's psychophysics dealt only with the relationship between physiological stimuli and their psychological perception. Wundt, however, wanted to develop an empirical science of purely psychological processes.

Consciousness, he holds, is the process of forming representations, which are initially compounded out of sensations, and the continuous flow and merging of these representations. Wundt argues that these associational processes go on and bring about the merger and combination of representations.

> Association everywhere gives the first impetus to [apperceptive] combinations. Through association we combine, e.g., the representations of a tower and of a church.[58] But no matter how familiar the coexistence of these representations may be, mere association does not help us form the representation of a church-tower. For this latter representation does not contain the two constitutive representations in a merely external coexistence; rather, in the [representation of the church-tower], the representation of the church has come to adhere [*anhaften*] to the representation of the tower, more closely determining the latter. In this way, the *agglutination of representations* forms the first level of apperceptive combination.[59] (*PP* II: 476)[9]

While these associational processes can go on in background, Wundt believes that actually paying attention to a set of representations requires an act of will. At any given moment, most of these representations will be in the background, but some will be the focus of consciousness. Wundt attributes the focusing of attention from one set of representations to another an act of will.

Wundt's methodology, which involves introspection or self-report, is comparative:

Wundt sees WL (Weber's Law) as simply a mathematical description of the more general experience that "we possess in our consciousness no absolute, but merely a relative measure of the intensity of the conditions [*Zustände*] obtaining in it, and that we therefore measure in each case one condition against another, with which we are obliged in the first place to compare it" (*PP* I: 393).[9]

While Weber, Fechner, Helmholtz and Wundt were forming the foundations of psychology, others were constructing the new discipline of sociology. August Comte (1798-1857), following in the footsteps of his mentor St. Simone, advocated the scientific study of society, which he called *sociology*. But Comte's understanding of science was negligible, and his notion of positivism bore no relation to the practices of the Ionians.

Opposition to the idea that society should be studied scientifically was swift: Karl Marx (1818-1883) argued that the correct method of studying human society was not science, but the dialectic, although he reversed Hegel's notion of dialectical idealism to his own notion of dialectical materialism. Max Weber (1864-1920) opposed the scientific study of society, preferring his own method of "verstehen" (German for "understanding") which was an interpretive method.

Emile Durkheim (1858-1917) was a strong proponent of the scientific study of society, but had little in the way of scientific education. He studied under Numa Denis Fustel de Coulanges, who was a classicist, and read August Comte and Herbert Spencer, neither of whom had any formal training in science in their background. But he did spend a year studying with Wilhelm Wundt, where he learned of Wundt's concept of *representations*, and learned Wundt's comparative methodology. Following Wundt, Durkheim's model of *collective consciousness* was a comparative concept that could be measured by the comparative method. Durkheim was the primary conduit of Wundt's work into sociology, although few sociologists know or acknowledge the influence of Wundt due to the compartmentalization of the social sciences.

Durkheim adopted Wundt's notion that consciousness was a process of forming representations from sensory stimuli, but which, through a subliminal process of associations, combined and agglomerated to become dissociated from their original sensory substrate. But Durkheim believed that concepts were formed not by individuals, but by collections of people, and these *collective representations* or *social facts* were the driving force behind the behavior of members of the society. These could be measured statistically, by rates of behavior, such as the number of births, deaths, marriages, suicides, and the like. In particular, he held that the average of such numbers represented the social facts with considerable exactness.[10]

Durkheim's understanding that the collective society preceded and determined the thoughts and behaviors of its individual members was a substantial departure from the Aristotelian individualism that had dominated Western philosophical thought until his time. Further, Durkheim's discussions of the movements of social facts relative to each other makes implicit reference to a spatial model invoking the Ionian concepts of distance and time, and although Durkheim never explicitly described the collective representations as mathematical points arrayed in space, later writers specifically referred to Durkheim's concept in their own spatial model of cultural processes.[11] While Durkheim's ideas would go on to have great impact on the newly forming discipline of Sociology, they would go largely unnoticed by psychology.

Wundt's work was introduced into the United States by Edward Titchener (1867-1927), who studied with Wundt for two years at Heidelberg. Unfortunately, Titchener's translations of Wundt's work from German to English distorted Wundt's model to make it more supportive of Titchener's own structuralist model. Titchener thought that it was possible to discover the internal structure of the mind through introspection, and his translations implied that Wundt supported this view, which he did not[7]. He began what was at the time the largest psychology department in the US at Cornell University, and died in Ithaca in 1927. His structuralist model did not survive his death.

---

[7] Wundt thought of the mind as a process, not a structure, and a structural understanding of the mind as Wundt thought of it is not possible.

Titchener was one of many scholars who, like August Comte, had a commitment to science, but did not really understand how it was done. Other founders of psychology were even less committed to science. G. Stanley Hall (1844-1924) was trained in theology before receiving the first doctoral degree in psychology from William James at Harvard. After a brief period at Wundt's laboratory in Heidelberg, he taught English and philosophy until securing a position teaching psychology in the philosophy department at John Hopkins. There he founded a psychology laboratory, where he advocated social reforms. He believed that people were basically ignorant, and needed to be organized by a strong leader. He recommended separating boys and girls during their early education so that men could be prepared for careers while girls could be trained to motherhood and domestic service. He was a champion of Sigmund Freud and Carl Jung, and arranged to bring them to the US to lecture.

Others had a stronger attachment to science. One of those was Joseph Jastrow (1863-1944) who was a fellow at John Hopkins where he met C. S. Pierce and collaborated on the design of experiments in psychophysics which were very advanced for their time, including randomization, blinding and repeated measures. He later was appointed chair of the psychology laboratory at the University of Wisconsin department of psychology. In 1918, his student Clark L. Hull received his PhD at Wisconsin, where he taught until 1929, after which he moved to Yale University, where he worked until his demise in 1951.

Apart from his brief collaboration with Pierce, neither Jastrow nor Hull had any direct connection to the Ionian network of scientists, but, as was the spirit of the times, favored a scientific approach. The result for Hull was a thinly disguised Aristotelian model of behavior, the thin disguise being expressions that resembled equations, e.g.,

sEr = sHr * D,
where
sEr = excitatory potential,
sHr = habit strength, and
D = drive strength.

Underneath the pseudo equations, however, lies Aristotle's model of appetites and goals.

Although Hull provided no way to measure the variables in the equation, he nonetheless elaborated it to include more and more terms, but there is a fundamental difference between the way Hull checks his theory against data and the way mature science does. To understand this difference, it's instructive to listen to Richard Feynman talk about the differences between how early scientists like Newton and contemporary science deal with discrepancies between theory and observation:

> Newton proposed this idea: Light striking the first surface sets off a kind of wave or field that travels along with the light and predisposes it to reflect or not reflect off the second surface.

He calls this process "fits of easy reflection or easy transmission" that occur in cycles, depending on the thickness of the glass.

There are two difficulties with this idea: the first is the effect of additional surfaces -- each new surface affects the reflection -- which I described in the text. The other problem is that light certainly reflects off a lake, which doesn't have a second surface. In the case of single surfaces, Newton said that the light had a predisposition to reflect. Can we have a theory in which the light knows what kind of surface it is hitting, and whether it is the only surface?

Newton didn't emphasize these difficulties with his theory of "fits of reflection and transmission," even though it is clear that he knew his theory was not satisfactory. In Newton's time, difficulties with a theory were dealt with briefly and glossed over -- a different style from what we are used to in science today, where we point out the places where our own theory doesn't fit the observations of experiment. I'm not trying to say anything against Newton; I just want to say something in favor of how we communicate with each other in science today.[12]

Feynman's position is quite clear: unless a theory correlates with observation *perfectly* (to within measurement error), it is wrong. But there is a different meaning to the word "correlate" in physical and social science that we will discuss in detail later. For now, however, it's enough to say that a theory correlates with experimental results in physical science if every observation conforms to predicted values to within measurement error, otherwise it does not correlate, and must be revised or rejected. In the social sciences, the theory is said to correlate with observations if measured values are roughly in the same direction as the theory predicts, and statistically "significant".

To understand this, it's useful to step back to Francis Galton (1822-1911) and his protégé, Karl Pearson (1857-1936). Galton and Pearson were eugenicists, or what we would today call racists, who believed that improvement of the human condition could only be brought about by interracial warfare in which the genetically superior races would wipe out the genetically inferior races, resulting in overall improvement for humanity. They believed that assistance to the poor, such as education, was wasteful, since advantages conveyed during one generation could not be genetically transmitted to subsequent generations, and so would have to go on perpetually.

Galton coined the phrase "nature or nurture," and devoted most of his research to trying to establish the truth of his claim that nature was responsible for whatever level of intelligence one might have.

Galton came to this opinion after reading his cousin Charles Darwin's *Origin of Species*. At that time, the notion that all men[8] were essentially equal was

---

[8] None of these thinkers would have granted woman the

commonly held. Thomas Hobbes (1588-1679) used the presumed equality of all men as the basis of his theory of government:

> NATURE hath made men so equal in the faculties of body and mind as that, though there be found one man sometimes manifestly stronger in body or of quicker mind than another, yet when all is reckoned together the difference between man and man is not so considerable as that one man can thereupon claim to himself any benefit to which another may not pretend as well as he. For as to the strength of body, the weakest has strength enough to kill the strongest, either by secret machination or by confederacy with others that are in the same danger with himself.
> And as to the faculties of the mind, setting aside the arts grounded upon words, and especially that skill of proceeding upon general and infallible rules, called science, which very few have and but in few things, as being not a native faculty born with us, nor attained, as prudence, while we look after somewhat else, I find yet a greater equality amongst men than that of strength. For prudence is but experience, which equal time equally bestows on all men in those things they equally apply themselves unto. That which may perhaps make such equality incredible is but a vain conceit of one's own

intelligence of men at this time.

wisdom, which almost all men think they have in a greater degree than the vulgar; that is, than all men but themselves, and a few others, whom by fame, or for concurring with themselves, they approve. For such is the nature of men that howsoever they may acknowledge many others to be more witty, or more eloquent or more learned, yet they will hardly believe there be many so wise as themselves; for they see their own wit at hand, and other men's at a distance. But this proveth rather that men are in that point equal, than unequal. For there is not ordinarily a greater sign of the equal distribution of anything than that every man is contented with his share.[13]

Through an amazing coincidence, the idea that all men were equal pervaded the enlightenment. (Of course it's not a coincidence; it's an idea that pervades the collective neural network which informs the individual brains of the people of the times.) By 1776, the US Declaration of Independence makes the same claim: "We hold these truths to be self-evident, that all men are created equal..."

The idea that all men are not created equal, and that some are genetically superior to others, was an idea that created the conflict that lead to the rise of fascism and Nazi Germany and hurled the world of the twentieth century into a great World War. It was a conflict that lay at the foundation of the social sciences, and an idea that dominated the thought of social and political thinkers for most of the twentieth century. It was a conflict that forged the fundamental methodologies and ideological struggles that lie at the root of the social sciences.

In order to prove their theory, Galton and Pearson developed a series of methods that today form the core of research methods in the social sciences. Among these were the questionnaire and the correlation coefficient.

Galton, Pearson and Walter Weldon founded the journal *Biometrika* in 1901 in order to advance their theories and new statistical procedures. In the first issue, they made clear what they expected to publish:

> *It is intended that* Biometrika *shall serve as a means not only of collecting or publishing under one title biological data of a kind not systematically collected or published elsewhere in any other periodical, but also of spreading a knowledge of such statistical theory as may be requisite for their scientific treatment.*

Pearson and Weldon co-edited the journal until Weldon's demise in 1906, and Pearson served as sole editor until his death in 1936, when he was succeeded by his son Egon, who served until 1966. The journal was a powerful force in popularizing the correlation coefficient in particular and statistical methods in general. In 1935, R. A. Fisher developed the concept of the *null hypothesis*, which Egon Pearson and Jerzey Neyman expanded and formalized shortly thereafter.

The protocol of the null hypothesis requires that any hypothesis put forward should be stated in null form -- "there is no relationship between a and b" and in alternate form "a is related to b." Random samples of cases are taken, where each case is a paired observation of a and b. Appropriate statistical tests are then made to calculate the probability that a and be are related; if this probability is higher than 90%[9] or 95%, the null hypothesis is said to be rejected. This does not prove, but supports the alternative hypothesis.

This notion of rejecting false hypotheses rather than accepting true hypotheses fits very well with the philosophy of Karl Popper, whose *Logik der Forschung* (titled in English *The Logic of Scientific Discovery*) was published in 1934. As we saw earlier, he argues that it is impossible to arrive at truth through induction, but insists that it is possible to produce increasingly useful theories by a rigorous process of rejecting false theories. Those theories that can survive rigorous falsification are more fit in a Darwinian way.

It is not a coincidence that these two very different lines of inquiry should converge on the notion of falsification to within a few months of each other. The rapid advance of Ionian science since Galileo had already overturned the most devoutly held theories of

---

[9] In theory, one should vary the level of probability of rejecting the null hypothesis based on the relative costs of making each of two kinds of mistakes -- rejecting a true hypothesis or failing to reject a false hypothesis, but in practice social scientists seldom use anything but the 90, 95 or 99 percent figures.

humanity, and done so repeatedly. The advent of relativity theory and quantum theory, showing that even the powerful Newtonian theory, after passing countless tests over centuries, was false, had made it increasingly clear that there was no way to prove a theory correct -- that the most satisfactory theory could be successfully challenged at any moment.

The Aristotelian concept of truth was unraveling, and the last intermediate stop on the way to abandoning truth altogether was the notion of falsification. By now many of the most advanced physical scientists had already abandoned the idea that their theoretical models were true or false, but instead were simply symbolic devices for organizing experiences in a way consistent with observations. Most social scientists, however, were a long way from abandoning the idea that they were converging on the truth. It was not yet clear to them that the idea of weeding out false theories until only true ones remain makes little sense if there is an infinite number of possible theories, particularly if the rejected theory leaves no clue as to how and by how much it was wrong.

We have recounted (superficially) in this book the history of physics from Thales to Feynman, and we have not seen even one instance of a test against the null hypothesis. The extent to which the combination of Popper's notion of falsification and Fisher's notion of the null hypothesis have been accepted by the community of social scientists is made emphatic by the fact that Wikipedia's entry on Null Hypothesis begins with this sentence: "The practice of science involves formulating and testing *hypotheses*, assertions that are capable of being proven false using a test of observed data."[14] *When social scientists are trained, they are led to believe that the Popper, Pearson, Fisher, and Neyman test against the null hypothesis is "the" scientific method.*

Popper's notion that better theories evolve through the rejection of false hypotheses requires that the theories be exposed to a *rigorous* process of rejecting false hypotheses. Now, as we shall see, in actual practice, testing against the null hypothesis has proven to be anything but rigorous, and this may well account for the fact that, in the half-century during which tests against the null hypothesis have been the required methodology in the social sciences, not a single theory has yet been rejected.

What we have seen in this chapter is the breaking of the chain of connections to the Ionian tradition. In Weber, Fechner, Helmholtz and Wundt we see Ionian scientists performing science using comparative methods. But by the middle of the twentieth century, almost all connections between the Ionian network of scientists and practitioners of social science are gone. The methods of research used by twentieth century social scientists are new, invented by statisticians and philosophers. Social science is cut adrift from science. Social science theories of human behavior fall back to Aristotle's *De Anima*, and are versions of Aristotle's teleological system in modern dress. Comparative measurement is the rare exception, and observations are made in categories. We are back in the fourth century B.C.E.

1. L. Patton, in *The Stanford Encyclopedia of Philosophy*, edited by E. N. Zalta (Stanford).

2. P. Pojman, in *The Stanford Encyclopedia of Philosophy*, edited by E. N. Zalta (Stanford, Stanford, 2011).

3. T. Uebel, in *Logical Empiricism -- Historical and Contemporary Perpectives*, edited by P. Parinni, W. Salmon and M. H. Salmon (University of Pittsburgh Press, Pittsburgh, 2003).

4. J. Fetzer, in *The Stanford Encyclopedia of Philosophy*, edited by E. N. Zalta (Stanfore University Press, Stanford, 2012).

5. T. Uebel, in *The Stanford Encyclopedia of Philosophy*, edited by E. N. Zalta (Stanford, Stanford, 2012).

6. S. Hook, *Psycholanalysis, Scientific Method, and Philosophy*. (New York University Press, 1959).

7. S. Thornton, in *The Stanford Encyclopedia of Philosophy*, edited by E. N. Zalta (Stanfore University Press, Stanford, 2011).

8. K. Popper, *Logic der Forschung*. (Julius Springer, Vienna, 1935).

9. A. Kim, in *Wilhelm Maximilian Wundt, Stanford Encyclopedia of Philosophy*, edited by E. N. Zalta (Stanford, 2008).

10. G. Simpson, *Emile Durkheim: Selections From His Work*. (Crowell, New York, 1963).

11.   J. Woelfel and E. L. Fink, *The measurement of communication processes: Galileo theory and method.* (Academic Press, New York, 1980).

12.   R. P. Feynman, *QED: The Strange Theory of Light and Matter.* (Princeton University Press, Princeton, New Jersey, 1985).

13.   T. Hobbes, *Leviathon.* (The University of Adelaide, South Australia, 2012).

14.   Anonymous, in *Null Hypothesis, Wikipedia* (2012).

Chapter 11: The Correlation Coefficient

"The correlation coefficient. The correlation coefficient! I hope to be rid of this noxious pest within my lifetime!" -- John Tukey

And now, social science veers sharply off the path of Ionian science by inventing new methods and procedures that will become the definition of science for them. Francis Galton and Karl Pearson invent a new method of calculating the relationships among variables that, they believe, does not depend on rational measurement.

Causality was a major casualty of relativity theory. Einstein's demonstration that the notion of simultaneity could not be established by observation, but rather was a matter of definition showed that even the temporal sequence of events was not invariant across change of reference frames, and so the simple notion of events being consequences of temporally prior events could not stand. If relativity theory wasn't enough to shake your faith in Aristotle's simple notion of efficient causality, quantum physics and Quantum Electrodynamic Theory (QED) provided the finishing blow.

Instead, scientists understood that their goal could not be to establish invariant sequences of cause and effect, but rather they must seek transformations that could establish correlations among the observations of multiple observers in multiple frames of reference. The meaning of "correlation" in this view, however, is not the vague idea that two or more sets of observations are related in some way. What it means is that transformations (e.g., rotations, translations, reflections, Lorentz contractions, etc.,) can be found which will establish *exact agreement* [10]among the observations of observers in multiple reference frames. Two observers' views *correlate* when they agree exactly on what they observe.

At about the same time, however, largely due to the work of Francis Galton and Karl Pearson, a new definition of "correlation" emerged. Pearson was an immensely influential polymath whose influence extended to many disciplines. He studied philosophy, science, mathematics, literature, biological sciences and others, and his ideas, particularly those expressed in his book *The Grammar of Science*[1] are known to have had some influence on Einstein's thinking. In this book, Pearson discussed the relativity of observations made in different reference frames, higher dimensional geometry, and causality, among other topics. His idealistic philosophy and his concept of relativity of motion across reference frames implied that causality was merely a human shorthand, and that all events in the universe can be thought of as correlated with all others, but that no causality exists. In all this, Pearson's

---

[10] As always, when we speak of "exact agreement" in science, we mean exact to within the uncertainties of our measurements, and subject to later revision based on further observations. Science is never perfect, and always fundamentally uncertain.

work can be seen to be clearly consistent with and perhaps influential over Einstein's later work.

It is not Pearson's philosophical or theoretical ideas that changed the definition of correlation for many who followed, but his practical invention of what he called the "correlation coefficient", or what we now technically call the "Pearson product-moment correlation."

Pearson's goal was to find a single number that would express the degree to which two "variables" were correlated[11]. His procedure is simple: first, write the variables as side-by-side columns of observations. Calculate the mean of each observation and subtract it from each value of the corresponding variable. Now both variables will be expressed as deviations from their means. If the variables correlate positively, in general when one variable has a value greater than its mean, the other will as well; when one variable has a value lower than its mean, so will the other. If the correlation is negative, when one variable has a value greater than its mean, the other is likely to have a value lower than its mean, and so on. That means that, if we multiply the corresponding values of the two variables and sum them up, those that are positively correlated will have a very high sum; those that are negatively

---

[11] Notice that even the idea of a statistic that describes how well two variables correlate is very different from the idea that transformation of events across reference frames must match exactly, or that theoretical predictions must conform to observations to within measurement error.

correlated will have a very high negative sum, and those that are completely uncorrelated will have a score of zero. Of course, the size of the sum of products will also depend on the number of products there are, but this problem is easily solved by dividing the sum of products by the number of cross-products (i.e., the sample size) and this yields the average sum of products, which Pearson calls the covariance, i.e., the extent to which the variables "co-vary."

But problems remain. The value of the sum of the products (covariance) will depend on other factors, such as the unit of measure of the scales on which the variables are measured. If a variable is measured in millimeters, for example, the sum of products will be a thousand times larger than if it were measured in meters. This is not a big problem, because we know how to convert millimeters to meters.

In the social sciences, however, the problem is more severe; if age is measured in years and propensity to view television is measured on a five-point scale, the sum of products is obviously meaningless in that it can't be compared to other sums of products for other variables measured on other scales, such as, for example, the sum of products of gender measured on a 0-1 scale and preference for ice cream measured on a 1-10 scale.

The reason this problem is so important in the social sciences is that many -- perhaps even most -- social scientists do not believe that psychological and cultural variables can be measured in the normal way of physical measurement, that is, as a ratio to some standard. Pearson's alternative was to compare the covariation to what he believed to be a universal standard, the *standard deviation*, a concept he took from his mentor, Francis Galton[2].

The essence of any variable is, of course, that it can take on different values as opposed to constants, which always have the same value. Variability is a matter of degree, and we can ask how variable a variable is. The answer proposed by Galton and adopted by Pearson is what is technically the root mean squared (RMS) deviation, which is given by the square root of the sum of the squared deviations from the mean divided by the number of cases, hence "root mean squares". Because he intended this to serve as a standard against which other measures could be compared, Galton called this the *standard* deviation.[12]

So Pearson can compare the covariation to the standard deviations. In mathematics, compare generally means divide, so we can divide the covariation by the product of the standard deviations (of course there are two standard deviations, because there are two variables). The result will vary between 1 and -1, since if the covariation is equal to the variation, the result will be 1 or -1 if the relationship is negative. If, of course, the covariation is zero, the result will be zero. This quantity

---

[12] A good measure of the difference in cultural mindset between Pearson and the Ionian comparative model is the fact that, in the Ionian model, a standard whose magnitude changes in different circumstances is worse than useless, while Pearson's standard deviation is different for every variable measured.

is called the Pearson Product Moment Correlation, which varies between 1 and -1, and is generally believed to be a measure of the strength of the linear relationship between any two variables.

It is impossible to overestimate the central role the correlation coefficient plays in the social sciences. Although investigators understand as a matter of faith that correlation does not prove causality, statistically significant correlations are generally considered to support theories and hypotheses, and non-significant correlations are generally a bar to publication in establishment journals. Why, then, did so significant a mathematician and statistician as John Tukey consider the correlation coefficient a "noxious pest" that he hoped to eradicate?

A deeper look into the mathematics behind the correlation coefficient reveals some real problems. First, the correlation coefficient and the standard deviation are not new inventions, but already exist in other branches of science and mathematics with other names. As we have seen, the standard deviation is also known as the root mean squares deviation. In linear algebra, the RMS deviation is also known as the norm of a vector, which is also the length of a vector. In linear algebra, the covariance of two variable vectors is actually the scalar product or inner product of those two vectors. The scalar product is, in fact, the product of the lengths of the two vectors and the cosine of the angle between them. Since the standard deviation is the length of the variable vectors, dividing the covariance by the standard deviations cancels out the lengths, leaving only the cosine of the angle between the vectors. The correlation coefficient, therefore, is the cosine of the angle between two variables. Rather than a measure of the strength of the linear correlation between two variables, then, the correlation coefficient is simply the cosine of the angle between them. If we find that two variables have a correlation of .5, we really mean that those two vectors lie at an angle of 66.6 degrees of each other.

The practice of dividing variables by their standard deviation is called *standardizing*, and is meant to solve the problem of variables measured on incommensurate scales. What it actually is, however, is a classic case of throwing out the baby with the bathwater, since the process of standardization eliminates all lengths entirely, leaving only angles. In Pearson's correlation world, we can only know in what direction any object lies, but never how far it is from us or anything else.

Figure 3 shows the first two principle axes of the straight-line distances among selected US cities.

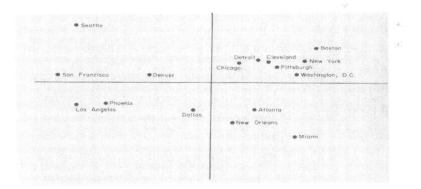

Figure 3: First Two Principle Axes of Distances Among Selected US Cities (from Woelfel & Fink, 1980)

Standardizing these data creates the grossly distorted picture shown in Figure 4:

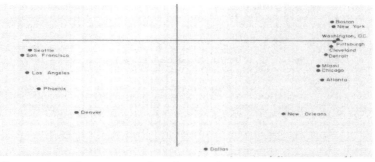

Figure 4:   First Principle Plane of Standardized Distances Among US Cities (from Woelfel & Fink, 1980)

Standardizing -- the process of expressing each value of a variable as a proportion of the length of that variable -- rather than solving the problem of incommensurate scales, introduces massive distortion even to well-measured data, creating such absurdities as Miami north of Chicago and Atlanta.

What all this shows is a serious change in the meaning of the word "correlation" from what Einstein and others meant and what the modern social scientist understands. For Einstein, correlation means an *exact agreement* among observers as to sequences of events -- any disagreement means there is no correlation. As Richard Feynman says, "If it doesn't agree with experiment, it's wrong." To the modern social scientist, correlation means only that there is less than a five or ten percent chance that two or more variables are mutually perpendicular.

We may recall that Kepler, in spite of his strong inclination toward perfect figures, rejected the idea that the planets orbited the sun in circular orbits, even though the average error introduced by the circular orbits averaged under 2 degrees. Contrast that with the psychologist Charles Osgood's unwillingness to reject his own theory when it is very far from supported by experiment:

> Osgood (1974) reports research into the semantic structure of 27 different cultural groups. The results produced loadings in the .80 to .90 range on the evaluative dimension, loadings from the .40s to the .70s for the potency factor, and for the activity dimension, .30 to .70. From these results, Osgood concludes, "This is rather convincing evidence for the universality of the affective meaning system".[3]

In the semantic differential system, the loadings represent the Pearson correlation of the position vector of the objects with the dimensions of the semantic differential space. Osgood's best correlation (.9) explains 81 percent of the variance in the location of the objects, an error of 19 percent. The best correlation with the other two dimensions is .7, which explains 49 percent of the variance, leaving an error of 51percent. The worst correlations -- .3 and .405 -- explain nine percent of the variance, with an error of 91 percent, and sixteen percent of the variance with an error of 84 percent. Just how much error is needed before a social science theory is considered wrong, or, at the very least, in need of some modification?

One of Pearson's main points in The *Grammar of Science* is that all experiences are connected. Indeed, it is very difficult to find two variables that are completely uncorrelated. Yet a standard practice in the social sciences is for a theory to predict a relationship between two (or more) variables, make observations, then compute the correlation between them. If the correlation is not statistically significant, the hypothesis is supposed to be rejected. But the level of statistical significance of the correlation coefficient is a function of the magnitude of the correlation, and the sample size. If the sample size is large (say, over 100 cases), very small correlations are statistically significant.

I once saw a colleague with a very large conference table covered with computer printouts. They were a large matrix of correlations among several hundred variables that had something to do with television viewing, such as the number of viewing hours per week, attitudes toward viewing, and the like. He was going over the correlations with a yardstick and yellow magic marker, highlighting the correlations that were statistically significant.

Since his research was well funded, he had a large sample size, and correlations as low as .15 were statistically significant. As he was taught in his methods classes, he believed this meant that these variables were significantly correlated and his theory was supported.

But a correlation of .15 shows that the angle between the two vectors is about 81 degrees, and explains a little more than 2% of their mutual variance. It's hard to find two variables that are less related than this, yet strictly following the mandatory rules of data analysis, he is unable to reject the null hypothesis, and considers his theory supported.

The theory my friend was failing to reject (we are taught in statistics class that we either reject or fail to reject the null hypothesis) was the dreaded Uses and Gratifications Theory.[4] With the widespread deployment of mass media -- by the 1930s these would be newspapers and other print media, radio and movies -- early Communication theorists worried that helpless individuals would be "massified" by the powerful one-way flows of information from the media. Gradually, opposition to this view grew up around the notion that individuals actively chose the media they used on the basis of the gratifications they expected to receive from them. This -- like Maslow's pyramid of needs -- is another thinly disguised version of Aristotle's theory, and one that Maslow actually claimed was a subset of his own theory.

Thanks to the support of moguls whose wealth derived from the mass communication industry like Roy H. Park and Walter Annenberg, mass communication research has historically been better funded than most communication research, Uses and Gratification supporters (like my friend) usually have large sample sizes, and their correlations are usually statistically significant. They are also remarkably small; in a review of recent Uses and Gratifications research, Cheong et. al., found explained variance ranged from a high of 27.2% to a low of 4%,[5] which corresponds to error from a low of 72.8% to a high of 96%, yet the theory is not rejected.

Once again, proponents of the theory continue to hold to it in spite of the very large error because they believe that human phenomena cannot be specified to much higher tolerances than these. But that is simply wrong.

Barnett et. al tested a model of television viewing that suggested that rates of viewing could be explained by seasonal factors such as the amount of daylight per day, by the diffusion curve for innovations, and by time independent factors. Television viewing data came from monthly Nielson reports for 460 months from 1950 to 1988. To calculate the frequency of the seasonality component, they used fast Fourier techniques which John Tukey had developed. Compare their results with the concept that human behavior is not precisely predictable, and with the dreadful track record of the Uses and Gratifications theory:

> The descriptive data for each variable are provided in Table1. The results of the test of the three-component model are presented in Table 2. The model fits the data very well, explaining about 99.9% of the variance in the frequency of television viewing. All of the model's parameters are statistically significant. The oscillation coefficients are plausible (i.e., they are nearly equivalent to the theoretically derived values), and the standard errors are quite small. Because

the estimate for B2 matches its theoretical value, it may be taken as a confirmation of the 12-month cycle. The coefficients for the translation component are also likely, because B4 (the maximum level of adoption) plus B7 sum to a value very close to the observed maximum. Further, the standard errors are small relative to the size of the coefficients. The time-independent term is also as predicted, and its standard error is quite small. An analysis of the residuals revealed that they are approximately normally distributed. The skew is 0.003 and kurtosis is -0.055. There are 229 cases with residuals greater than 0 and 231 with negative residuals. There are 20 cases (4.4%) greater than 2 standard deviations of the residuals. Eight are negative, and 12 are positive. The total is less than what one would expect by chance (23). The hypothesis that the residuals come from a normal population cannot be rejected (p < .15).[6]

By now it should be clear that proponents of the Uses and Gratification theory don't hold to the theory because of any scientific evidence. They adhere to it because the belief that human beings are inherently goal seeking animals who choose freely among alternative behaviors based on a calculus of potential rewards and costs is at the very core of the culture to which they belong. The tests against the null hypotheses, the Likert scales, the scholarly articles and all the rest are simply rituals supporting the culture.

This is hardly the "rigorous process of falsification" Popper had in mind, and a major reason why no theory put forward by social science has ever been rejected.

1. K. Pearson, *The Grammar of Science*. (Black, London, 1892).
2. F. Galton, *Hereditary Genius*. (Macmillan, London, 1869).
3. G. A. Barnett, R. Wigand, R. Harrison, J. Woelfel and A. Cohen, Human Organization 40 (4), 330-337 (1981).
4. E. Katz, J. G. Blumler and M. Gurevitch, Pubic Opinion Quarterly **37**, 508-523 (1973).
5. P. Cheong, J. Hwang, B. Elbirt, H. Chen, C. Evans and J. Woelfel, in *The interrelationship of business and communication*, edited by M. Hinner (Peter Lang, Berlin, 2010), Vol. 6.
6. G. A. Barnett, H. Chang, E. L. Fink and W. Richards, Communication Research **18** (6), 755-772   (1991).

Chapter 12: Through a Glass, Darkly

Darwin's Descent of Man upset the Enlightenment belief that all men were equal. Spearman and Thurstone developed mathematical procedures to try to discover what "factors" underlay intelligence. In doing so, they created mathematical models of space that were quite strange...

Charles Spearman (1863-1945) was a soldier in the British army when he developed an interest in experimental psychology. Since psychology was still viewed as a part of philosophy in Britain, he traveled to Germany to study with Wilhelm Wundt. As we've seen, after a very long period during the Enlightenment when it was commonly believed that all men were essentially equal in intelligence, Darwin's origin of species, particularly as interpreted by Francis Galton, had generated intense interest in human intelligence. Spearman, who was strongly influenced by Galton, was determined to devise a precise scientific measurement of human intelligence. Galton had tried to show that intelligence as measured by class grades and ranks was correlated with physiological characteristics such as reaction time and the like, but could find no such correlations.

Spearman's principle method in this was to take the matrix of Pearson product moment correlations among students' scores on a battery of tests and decompose[13] them into a smaller set of vectors he believed represented the basic factors underlying intelligence -- which is why he named his procedure "factor analysis." Spearman's procedure generally produced one large "factor" and a number of smaller factors. He considered the large factor to represent general intelligence, and referred to it as the g factor, while holding that the several smaller factors represented special abilities.

Louis Leon Thurstone (1887-1955) began his academic career as a student of electrical engineering at Cornell University. It was there that he first developed an interest in psychology. He was interested in photography, and invented a movie camera that eliminated flicker by means of a continuously moving film. Thomas Edison was interested in the camera, but more interested in Thurstone, whom he hired to work in his laboratory after his graduation from Cornell. But Thurstone wanted to return to academia, and left Edison after two years to take a position as instructor in engineering at the University of Minnesota, where he also took a course in experimental psychology. In 1914, he enrolled in the graduate school of psychology at the University of Chicago, and accepted an assistantship at the Carnegie Institute in 1915. He earned his doctorate from Chicago in 1917, and soon became a professor and head of the psychology department at Carnegie in 1920,

---

[13] The "factors" were actually the eigenvectors of the correlation matrix, which, as we have seen, is itself a (standardized) scalar products matrix.

a position he held until 1923. In 1924 he moved to the University of Chicago, where he remained until his retirement in 1952. He then became Research Professor and head of the psychometric laboratory -- later renamed in his honor -- at the University of North Carolina, a post he held until his demise.[1]

Thurstone strove to establish psychology as a scientific discipline. He saw limitations in the work of Weber and Fechner because their research applied only to phenomena that had a physical manifestation, such as light or sound or pressure, and hoped to develop measures of purely psychological variables, such as attitudes and intelligence.

In the case of intelligence, Thurstone, like Spearman, gathered matrices of correlations among batteries of tests and factor analyzed them (a truly arduous task requiring many research assistants before the computer!), but he disagreed with Spearman's analysis. Thurstone thought that the principle axes might not be the axes that represented the individual factors, and suggested rotating the matrices until certain criteria were maximized or minimized in a search for what is called "simple structure". This led to a series of criteria such as varimax, minimax, quartimax and other mathematical criteria, each yielding different results. Thurstone also suggested the possibility that abilities might not be independent of one another, so that oblique[14] rather than orthogonal rotation schemes

---

[14] As we've seen, variables that are not correlated with

came into being. Thurstone disagreed with Spearman and insisted that there were seven factors underlying intelligence.

After Thurstone, factor analysts developed more and more *ad hoc* procedures, such as eliminating all factor loadings under .7, or .4. Because the mathematics were so abstract and difficult, and because so many possible alternative ways to carry out factor analysis existed without clear, undisputed reasons to choose one over others, factor analysis, although still used in social science, remains murky to most social scientists, even those who use it. And because of the large range of alternative ways to do it, one can move among alternating methods until achieving the result one hopes for.

This is particularly true in the social sciences, because social scientists do not have the respect for observation that underlies the physical sciences. If you believe your observations are suspect, you may feel free to change them or ignore them completely. (Many psychometric procedures involve altering or disregarding measured values.) Underlying all this, of course, is the fact that most factor analysis begins with a matrix of correlation coefficients, which we know to be

---

each other are perpendicular to each other. Since Thurstone believes that the various mental abilities are correlated with each other, he must allow them to be "oblique", i.e., not at right angles from each other. Bear in mind also that the meaning of the word "correlate" is taken here in the weak Pearsonian sense, not in the strong Einsteinian meaning.

the cosines of angles between variables, so vast distortion is built in even before the analysis begins. Disregarding the lengths of the measured values (standardizing) is, of course, what the correlation coefficient is all about.

Factor analysis represents a kind of space. Since Spearman was interesting in finding the "factors" underlying intelligence, he called the dimensions of the space "factors", but they are dimensions nonetheless, and the "loadings" of items on the factors are their coordinate values. The point which lies at the intersection of all the "loadings" or an item on the various factors is the location of that item in the space. Since correlations are standardized to have unit length, every item lies exactly 1 unit away from the center of the space, and so all items lie on the surface of a unit hypersphere -- a multidimensional sphere one unit in radius. Inside the hypersphere is nothing, since everything lies on its surface.

What is interesting in this context, however, is that the concept of space underlying both Spearman's and Thurstone's understanding of factor analysis is a relatively primitive concept of absolute, not relative, space, in which different regions have different properties. Up is different from down, left is different from right. Its origin is a special place, and all mental attributes are lines passing through the origin. The dimensions themselves are thought to be meaningful -- a far cry from the space of Newton and Einstein. Although neither Spearman nor Thurstone claim that fire belongs at the periphery of the space, or that heavy objects belong at the center, they do claim that smart people belong at the positive end of the x axis, and dumber ones at the left end. It is also a static space in which motion is not defined. As Spearman says "G is in the normal course of events determined innately; a person can no more be trained to have it in higher degree than he can be trained to be taller. "[2]

The work of Spearman and Thurstone, along with that of Cattel, Binet and others working in the area of human intelligence measurement, is testimony to the power of a cultural preconception. Prior to Darwin's *Descent of Man*, it was the universal belief of Enlightenment scholars that "all men were created equal" in their mental abilities. Now, however, even the brightest lights in science are overcome by the notion that some men -- and some races -- are more intelligent than others, that nothing can be done about it, and it is morally correct for the brighter to eradicate the less intelligent through any means at all. And now, some of the best and brightest believe it is possible to come up with a single number -- or even seven numbers -- that measure the fixed, unchangeable ability of a human being.

The idea that human ability is fixed defies everyday observation. Non-musical children who take piano lessons are suddenly able to play the piano. Prior to Newton and Leibniz inventing the calculus, the greatest geniuses on earth were unable to describe complex motions and curves with precision; afterwards, ordinary high school students could do so. A classic example of how learning new symbols increases mental ability is factor analysis itself. Without a grounding in linear algebra, factor analysis is obscure beyond comprehension for even the brightest people. But a little linear algebra makes it easy to understand. In fact, the major argument of this book is that the Ionian culture makes ordinary people able to do extraordinary things, while other excellent -- even extraordinary -- people, hobnobbed by the Athenian model, are unable to make simple observations.

The brain, at birth, is not yet complete, nor will it ever be complete until death. As a result of natural selection, the newborn brain is well suited for breathing, circulating blood, maintaining a constant temperature and other physiological functions, but it lacks the circuitry it will need to recognize its mother, to talk, to play the violin. The way in which this learned circuitry is wired has profound effects on the brain's abilities in every regard.

Of course, it may well be that Spearman and Thurstone intended to measure the abstract capability of a newborn brain to accept new wiring, but there is no way one can measure this by means of questionnaires or tests that rely on the learned circuitry even to read the test. It would be foolish to claim that we will never learn how to measure such an innate capacity, but we certainly are not able to do that now.

Whatever its merits as a measure of mental ability, factor analysis marks one of the earliest concepts of space in the social sciences.

1. J. P. Guilford, (American Academy of Science, 1957).
2. Anonymous, *Charles Spearman, Wikipedia* (2012).
3. A. Ferguson, Myers, C.S., Bartlett, R. J., Banister, H., Bartlett, F. C., Brwn, W., Campbell, N. R., Craik, K. J. W., Dreyer, J., Guild, J., Houstoun, R. A., Irwin, J. O., Kaye, G. W. C., Philpott, S. J. F., Richardson, L. F., Shaxby, J. H., Smith, T., Thouless, R. H., Tucker, W. S., Advancement of Science **1**, 18 (1940).
4. R. D. Luce and J. W. Tukey, Journal of Mathematical Psychology **1**, 27 (1964).
5. Anonymous, *Level of measurement, Wikipedia* (2012).
6. D. B. S. Shinde, *Research Method in Physical Education.* (Sports Publication, New Delhi, 2011).
7. S. S. Stevens, Science **103** (2684), 4 (1946).
8. R. D. Luce, British Journal of Psychology **88**, 4 (1997).
9. R. N. Shepard, Psychometrika **27** (2), 15 (1962).
10. J. B. Kruskal, Psychometrika **29**, 14 (1964).
11. G. Rasch, *Probabalisitc models for some intelligence and attainment tests.* . (Danish Institute for Educational Research, Copenhagen, 1960.

Chapter 13: Blinded by the Likert

And now, social scientists conclude that human phenomena cannot be measured in the traditional way of science, and invent not only new kinds of measurement, but a whole new theory of measurement: the nominal, ordinal, interval and ratio scales. Physical scientists don't recognize the first two as measurement at all, but these are virtually the only ones social scientists use.

Thurstone's second quest was the scientific measurement of human attitudes. Thurstone envisioned a space in which were arrayed various "positions" people could take toward any issue. By looking at the magazines available at a magazine stand, for example, one could find material about issues of the day -- say, about labor unions. Of course there would be differences of opinion about these issues, and each of these positions could be collected and written on a card. One card might say "Labor unions are vitally necessary to ensure the rights of workers." Another might say "Labor unions violate free trade and damage the economy." Thurstone then hoped to be able to select from this sample of (perhaps 100 or so) positions a smaller set such that each position was roughly equidistant from its neighbors, beginning with the most unfavorable positions and arrayed in ascending order until the most favorable position.

This he accomplished by having a sample of people sort the position cards into eleven piles from least favorable to most favorable. After each person had sorted the card, the number of the pile into which it had been sorted (1-11) would be written on the back of the card. After a large number of people had sorted the cards, he would calculate the average of all the numbers on the back of each card and calculate the variance around the mean (generally but not necessarily the interquartile range.)

If the variance is high, the item is rejected, since high variance indicates the judges could not agree on what pile the position belonged in. The remaining set could be used as a questionnaire, and the position of the respondent on the array of the positions would be given by which items he or she agreed with. The person would lie in the space close to the positions he or she agreed with. Like Factor Analysis, in which Thurstone also played a large role, this is an early concept of space in the social sciences.

The amount of effort involved in the construction of a Thurstone scale is not at all extreme when compared to the efforts involved in physical science research -- compare it for example to the effort expended on the Michelson-Morley experiment -- but it is much more tedium than social scientists who do survey research are accustomed to.

Thurstone's scale allows respondents only a dichotomous choice, agree or disagree, for each statement in the questionnaire. In an effort to improve Thurstone's scale, Rensis Likert (1903-1981) allowed a range of agreement, typically from strongly disagree to strongly agree in five steps, with the middle point neutral. While neither the Thurstone scale nor the Likert scale would be recognized as genuine measurement in physical science, which uses only comparative measures, Likert and others were convinced his five point scales provided extra information for the scale and represented an improvement over the Thurstone scale.

Whether either scale provides useful measurement becomes moot, however, since the Pearson product moment correlation between the five point items and the entire summated scale are quite high. Although we've already discussed some of the problems with the correlation coefficient, social scientists embraced Pearson's r with unusual unanimity, and began to use the five point Likert type scales *instead* of the summated scales both Thurstone and Likert had intended. Today, in practice, virtually no one uses either a Thurstone Scale or a Likert scale, but the little five-point devices (often called Likert-type scales, or, erroneously, Likert scales) have become ubiquitous, being perhaps the most widely used psychometric instrument in all of social science. This would be a serious problem for astronomers, who would have to be clever indeed to figure out how to measure the distances among the eight planets and the sun on a five point scale.

The Pearson correlation between the actual mean distances of the planets from the sun and their corresponding Likert distances is .64, which most social scientists would consider quite good. That could result in a huge savings for NASA. In Likert world, Earth is 3 units from the sun, while Mars is 4, so they will only have to fly 1 unit instead of 160 million miles to get the rover up there. It would get a bit murky in the middle, though, when the spacecraft crossed over from 3 (earth's position) to 4 (mars' position), not unlike Aristotle's theory of motion. Telemetry about the spacecraft's position would be uninteresting, as "...3,3,3,3,3,3,3,4,4,4,4,4,4..." Clearly, Likert's procedure has lost the sense of space that underlay Thurstone's model, and has become categorical rather than comparative.

Meanwhile, Stanley Smith Stevens (1906-1973) continued the work of Weber and Fechner. Stevens measured the physical intensity level of various stimuli (sound levels, intensity of light, pressure and the like) and asked respondents to measure the intensity of their perception of those stimuli as ratios to a standard stimulus. In the social sciences this is referred to as *magnitude estimation scaling*. As it is understood in the physical sciences, it is comparison to some standard.

While in physical science, measurement always means comparison to a standard, in the social sciences, measurement means assigning numbers to observation according to some rule, and there are thousands of alternative rules. The fact that measurements made by different investigators in different areas for different topics at different times based on different rules were incommensurate was widely understood early on, and led to the formation of a committee established in 1932 by the British Association for the Advancement of Science to investigate the possibility of genuine scientific measurement in the psychological and behavioral sciences.

This committee, which was known as the *Ferguson committee*, published a Final Report[1] which was highly critical of measurement in the social and behavioral sciences. One of the members, Norman R. Campbell, argued strongly that Stevens' measurements were not only false but meaningless unless and until he could establish that they could be shown to be *additive*, that is, the simple operation of addition was possible. Later writers[2] showed Campbell's criticism to be invalid, but Stevens chose not to respond to the specific argument, but rather to present an entirely new theory of measurement[15].

In a profoundly influential article in *Science*,[5] Stevens made the remarkable ontological argument that *there existed* four classes of entities which could be distinguished on the basis of their measurability. Some things could be measured comparatively, as ratios to a standard. Others could be arrayed on the number line such that the distances among them could be precisely

---

[15] Two remarkably similar accounts of these events can be found on Wikipedia: 3. Anonymous, in *Level of measurement, Wikipedia* (2012). and a book on research methods. 4. D. B. S. Shinde, *Research Method in Physical Education*. (Sports Publication, New Delhi, 2011).

measured, but they could not be compared as ratios because of a lack of a true zero point. But Stevens proposed two more quite amazing classes of entities, those which could be arrayed in order[16], but distances among them could not be established, and finally, entities that could only be named.

Today, few psychometricians -- those who study measurement professionally --lend any credence to Stevens' theory of measurement, but the world of psychometricians is for the most part inaccessible to the working social scientist. Stevens' fourfold classification of measurement is a central part of every introductory social science textbook on research methods, and the idea that there are nominal, ordinal, interval and ratio scales, and that each is appropriate to some kinds of

---

[16] At first blush it may seem easier to establish the rank order of items than to measure their actual magnitudes -- you can easily notice that you prefer vanilla to chocolate without calculating how much you prefer it. But as the number of items to be ordered grows even moderately large, the number of comparative operations needed to establish their rank order grows very rapidly (to rank order N things requires N(N-1)/2 comparisons), while the number of comparisons needed to establish the magnitudes of a number of items compared to some standard is always equal to the number of items (N). It's fascinating to realize that a whole methodological area -- non parametric statistics -- has grown up around the *a priori* assumption that there exists a class of entities that can only be placed in rank order -- an assumption supported only by Steven's 1946 assertion.

phenomena, is an article of unshakable faith in the social science community.

Most social scientists, following Stevens, believe that different variables have inherent characteristics that affect their "measurability". Some psychometricians, such as Luce, have grave reservations about Stevens' model, but still believe in the existential character of variables and provide mathematical adjustments based on the distributional properties of observations to yield results that meet mathematical assumptions they believe would characterize "correct" measures.[6]

The idea that knowledge is perfect and unchanging and absolute is deeply ingrained in the global neural network that makes up the collective consciousness of social science, as is the idea that this world of experience is flawed and not the source of true knowledge. Most social scientists continue to believe that observations made of this world of experience are not the true object of our quest, but that, through an appropriate selection of assumptions, it is possible to "adjust" the measured values of observations to reveal the "true nature" of the variable measured[17].

Some of the most talented psychometricians, for example, make assumptions about what the underlying geometry of a given configuration of data points "ought to be", and adjust measured values until they produce the "expected" geometry. Roger Shepard[7] and Joseph

---

[17] Platonism is an occupational hazard of the mathematician.

Kruskal[8] independently developed clever mathematical procedures called *nonmetric multidimensional scaling* for transforming data which exhibits non-Euclidean properties into fully Euclidean data. However ingenious mathematically, these procedures are based entirely on two *a priori* assumptions: *first*, that psychological phenomena cannot be measured accurately, so measured data are not trustworthy, and *second*, the space within which psychological processes are arrayed is Euclidean and of low (two or three) dimensionality. Similarly, proponents of Rasch measurement make assumptions about how a perfect scale ought to be distributed, and adjust measured values until those criteria are met.[9]

The perspective or paradigm underpinning the Rasch model is distinctly different from the perspective underpinning statistical modeling. Models are most often used with the intention of describing a set of data. Parameters are modified and accepted or rejected based on how well they fit the data. In contrast, when the Rasch model is employed, the objective is to obtain data which fit the model. [10]

These differences in belief between physical and social scientists lead to irreconcilable differences in the definition of measurement. For physical science, the definition of measurement is clear and simple: measurement is *comparison to some standard.* For the social sciences, with their complex ontological assumptions about the measurability of different phenomena, the definition is much more complex:

measurement is *the assignment of numbers to observations according to some rule.*[5]

While controversy about the status of measurement of cognitive and cultural variables may exist in some arenas, at the level of the introductory textbook, the argument has been long settled. Social scientists in all disciplines are taught from the beginning that *there are four kinds of variables, and that the principle variables of the social sciences can only be measured at a level somewhere between the ordinal and the interval, that five point scales are appropriate for most measurements, and that relations among variables can be described with sufficient accuracy by the cosine of the angles between them.* And, as we learned earlier, *all inferences are only probable, but a one and ten or one in twenty probability (.90 or .95) is enough to support a scientific decision.*

The difference between physical and social science can be stated very clearly. In social science, a theory claims that two or more variables are related, the values of the variables can be measured with five point scales, and the correlation coefficients among the variables can be calculated. If the correlations are different from zero with a likelihood better than one chance in ten or twenty, the theory cannot be rejected and, in practice, is considered to be supported. In physical science, a theory claims that two or more variables are related in a specific functional way. Measurements are made as comparisons to a standard, and results must correspond to theoretical predictions to within the uncertainties around the measurements. If all the measurements don't conform to predictions within these tolerances, the theory is wrong. If they do, measurements of increasing precision are made and the cycle is repeated.

It's easy to see that the test of theory against observation is vastly more stringent in the physical sciences than the social sciences. In fact, if we understand that the world is an organic, holistic place were virtually all variables are related to all other variables to some degree, however small, it is easy to see why no social science theories have ever been rejected by experimental evidence. When exposed to social science tests, none of the many theories proposed by social scientists are false. If exposed to tests meeting the criteria of physical science, they are probably all false.

1.   A. Ferguson, Myers, C.S., Bartlett, R. J., Banister, H., Bartlett, F. C., Brwn, W., Campbell, N. R., Craik, K. J. W., Dreyer, J., Guild, J., Houstoun, R. A., Irwin, J. O., Kaye, G. W. C., Philpott, S. J. F., Richardson, L. F., Shaxby, J. H., Smith, T., Thouless, R. H., Tucker, W. S., Advancement of Science **1**, 18 (1940).
2.   R. D. Luce and J. W. Tukey, Journal of Mathematical Psychology **1**, 27 (1964).
3.   Anonymous, in *Level of measurement, Wikipedia* (2012).
4.   D. B. S. Shinde, *Research Method in Physical Education*. (Sports Publication, New Delhi, 2011).

5. S. S. Stevens, Science **103** (2684), 4 (1946).
6. R. D. Luce, British Journal of Psychology **88**, 4 (1997).
7. R. N. Shepard, Psychometrika **27** (2), 15 (1962).
8. J. B. Kruskal, Psychometrika **29**, 14 (1964).
9. G. Rasch, *Probabalisitc models for some intelligence and attainment tests.* . (Danish Institute for Educational Research, Copenhagen, 1960).
10. Anonymous, in *Rasch Model, Wikipedia* (2012).

Chapter 14: Very Darkly

And now, the consequences of abandoning the Ionian scientific model of measurement begin to take their toll. No one notices, because they believe the confusion they see is natural to human phenomena. But applying social science measurement and analysis procedures to known physical problems shows their devastating effect.

Social science measurement procedure is very different from measurement in the physical sciences and engineering, and, in fact, by strict standards of scientific measurement, should not be considered measurement at all. While it is commonplace to note that reality is socially constructed, few social scientists understand that the Treaty of the Meter was the fundamental step in the process of constructing the concept of space and time as we understand it today. Before the Treaty, standard measures of time and distance were crude. In conducting his famous inclined plane experiment by which he arrived at the law of falling bodies, some commentators believed that Galileo had no clock sufficiently precise to measure the duration of the fall, so he used an inclined plane to slow the falling to a manageable rate, and then -- as some now suspect -- sang or chanted, keeping time to the music, as he measured off the distance the ball rolled.

Had he used the most common social science methods, however, the result would have been much different.

To test this assumption, Galileo's experiment was replicated with minor alterations:

A 48" long Formica-topped table was tilted slightly by placing 200 page paperback books under the legs at one end. A 3/4" plastic ball was released at the higher edge of the table and allowed to roll down the incline until it dropped off the other edge. The time of rolling was measured with a Hewlett-Packard HP55 timer, and the distance the ball had rolled at one-second intervals was marked off. Instead of using a comparison to a standard rule, however, the distance rolled was measured using the most common social science rule: a numbered category scale, where 1-very short, 2=short, 3=neither long nor short, 4= far, and 5=very far. Ten trials were run for a total of 50 measurements. The mean distance rolled was 2.76±.20. The power curve $s=at^b$ was estimated, with parameters $s=.96t^{.95}$, and fit with an r^2 of .908. Equally good (or bad) was the linear equation $s=a+b^t$, with parameters estimated as $s=-.02±.92^t$, which fit with an r^2 of .904. The worst fitting equation is the correct law, $s=a+bt^2$, whose parameters were estimated as $s=1.11+.149t^2$, and a fit of $r^2=.883$.

If this experimenter had persisted, and constructed a new, bigger apparatus, and rolled more balls, the situation would have become even more confusing, since the numbers 1 through 5 in the first experiment would have referred to different distances than the same numbers would have on the larger apparatus. If, for example, he built a table twice as long,

"5" would mean roughly between 76 and 96 inches, whereas in the original study, "5" meant roughly 38-48 inches. Clearly, if physicists used the most common social science measurement rule, they would not have arrived at the same "laws of nature" we now know. Instead, their world would contain an unpredictable component (the 10% unexplained variance in this experiment), and they might even have concluded that reality is not law governed, assigning the ball a measure of "free will."

It's not just the laws of nature that would be disrupted if physics were to adopt social science measurement procedures, but our entire picture of the world would be vastly different. Using the method of comparison to the standard meter -- the measurement rule for the physical sciences -- we are able to say that the diameter of the sun is 1.4 million kilometers, the diameter of the moon 3476 kilometers, the diameter of a US quarter 24.26 millimeters, and a US nickel 21.21 millimeters. We can say that the sun is about 66 billion times larger than a nickel.

If we apply the most common social science measurement rule, however, the world looks very different. On the face of it, the social reality that can be constructed out of five points scales is very crude, allowing for at maximum a ratio of five to one (i.e., something could be, at most, five times larger than something else) compared to the 66 billion to one ratio of the sun to a nickel. In practice, the distortion is even worse. Consider the following experiment:

Twenty-four undergraduate students were

randomly divided into three groups. Each answered four questions. Each question was collected before the next was administered. The questions were: How large is the sun? How large is the moon? How large is a quarter? How large is a nickel? Responses were 1=very small, 2=small, 3=neither large nor small, 4=large, 5=very large. The questions were administered in this order:

Group 1	Group 2	Group 3
Moon	Sun	Nickel
Sun	Moon	Quarter
Nickel	Nickel	Moon
Quarter	Quarter	Sun

Table 3: Order of presentation for three groups

The results of these measurements are:

	Group 1	Group 2	Group 3
Moon	3.38±1.41	4.00±.93	3.88±1.25
Sun	4.25±1.16	4.25±1.04	4.13±.99
Nickel	2.13±.64	2.25±.71	2.50±.53
Quarter	2.13±.64	2.38±.52	3.25±.71

Table 4: Size of Moon, Sun, Nickel and Quarter for three groups

Although the actual ratio of the size of the sun to the size of a nickel is about 66 billion to one, and the maximum theoretical ratio possible in a five-point scale is five, the actual ratios measured in the experiment are shown in the following table:

	Group 1	Group 2	Group 3
Moon/Sun	.80±1.82	.94±1.40	.94±.59
Moon/Nickel	2.0±1.72	1.89±1.07	1.65±1.36
Sun/Quarter	2.0±1.32	1.79±1.16	1.27±1.21
Nickel/Quarter	1.0±1.17	.95±.88	.77±.89

Table 5: Ratios of sizes of Moon, Sun, Nickel and Quarter for three groups

The largest ratio actually measured in Table 5 is 2.0, showing respondents estimated the Moon to be twice the size of a nickel, and the Sun to be twice as large as a quarter. Respondents report that the moon is between 80% and 94% of the size of the sun, and one group reports the sun is only 27% larger than a quarter.

Once again, however, the deeply ingrained notion of two worlds -- the world of unreliable experience and the real or genuine world -- arises to worsen the situation. In statistics, social scientists are taught that their observations are always constrained to samples drawn from a universe or population. Although the sample is the only thing we can observe, our real interest is in the (unobservable) population. We can, we are told, on the basis of some mathematical and statistical assumptions, *reason* to what the "real" population must be like. In the present case, using the standard rules of inference from sample to population, if standard 90 or 95 percent confidence intervals are used, we would conclude that these subjects show no significant differences in size between the Sun, the Moon, a quarter and a nickel. In the platonic "population", our research shows that the Sun, the Moon, a quarter and a nickel are exactly the same size.

The point of this exercise is not to show the low level of education of the typical undergraduate. These students may not know that the Sun is 66 billion times larger than a nickel, but they do know that it is vastly larger. The standard social science methodology, however, *prevents them from expressing this knowledge.*[1]

In the physical sciences, if a theory doesn't agree with observation, the theory is wrong. But not in social science -- or medical science, either. In the largest health related study of all time, researchers at Harvard University found beneficial effects of dietary fat reduction on heart disease, breast cancer and colon cancer in their 25 year sample of 161,000 women, but the result was not statistically significant, so they were "forced" to conclude that there was no beneficial effect in the population. The consternation in the investigator's words is apparent in this transcript from Talk of the Nation:

> The federally funded study, known as the Women's Health Initiative, tracked 161,000 women, lasted 15 years, and cost $725 million. It was designed to test different strategies for preventing heart disease, bone thinning osteoporosis, and breast and colorectal cancer. It made news in 2002 when its findings created doubt that hormone therapy was good for women's overall health. It again made news last month when final results did not conclusively show that a low-fat diet reduces the risk for heart disease, breast cancer, and colon cancer. It

also did not show whether or not calcium and vitamin D offer protection against broken bones.

CLARK: This morning I was listening to Diane Rehm Show and the doctors, previous leaders of the NIH, seem to say, if I got my information correct, that calcium along with vitamin D helped reduce hip fractures by twenty some odd, 26 to 29 percent. And then at the beginning of your show here, one of your comments to introduce the show was the saying that calcium, the study showed calcium may not matter, or didn't help. And if I had just heard that and then turned my radio off, I might call my mom and tell her don't worry about her taking the calcium.

Dr. HOWARD: Well, I'm going to defend the press there in that the distinction here is that, as trialists, there is a scientific way that one has to present the result. And first, you have to show the group as a whole, and the reduction in hip fractures in all the groups, all the ages, was not, it didn't reach that level of statistical significance. It was close again, but it didn't make it. So, we always have to say that first. And that's of course what the headlines picked up. But what he's pointing out is, as soon as you look to the next level you see for all the women over 60, there was a 21 percent reduction. And for the women who actually took their pills, because whenever you do a trial, and we had quite good adherence in that trial, in other words people taking their pills, but some didn't, if you took them out, then there was a 29 percent reduction in the whole group. So that's the problem. We have to report it, in a scientific article, the way it's expected for good science, and that then the news report is going to show the primary finding. And, if people don't read past the headline, they don't get the rest of it.

Dr. Howard's disclaimer on National Public Radio did little to ameliorate the public confusion when statistical significance testing compelled the researchers to disregard the obvious benefits of low fat diets on health and report that the results were not significant, and that, in the population, they couldn't reject the hypothesis that there was no relationship between dietary fat consumption and cancer and heart disease. Clearly the theory -- for statistical inference theory is indeed a theory -- outweighs the observations in the social sciences.

How could Harvard University -- the epitome of Establishment orthodoxy -- get it so wrong? John Tukey, whom we met in Chapter 11, the arch enemy of the correlation coefficient and developer of the Fast Fourier Transform that Barnett and his coworkers used to identify the factors that determined television viewing, spent his academic career 263 miles from Harvard in another bastion of establishment thinking, at Princeton. He was a brilliant mathematician and statistician, and a friend of Richard Feynman from graduate school.

Feynman and Tukey were both curious characters. Feynman had investigated his ability to count while carrying out other tasks, and found that he could do many things while counting, but he could not speak and count at the same time. Tukey, on the other hand, could talk and count simultaneously -- but he could not *read* and count. Feynman counted sub vocally, with his ears, but Tukey saw the numbers streaming past his eyes.[2]

In April, 1977, Peter Monge, a Professor at another pillar of the Establishment at the Annenberg School of Communication at the University of Southern California, arranged for a conference of the leading methodologists in the field of Communication at Asilomar, in Pacific Grove, California, sponsored by the Office of Naval Research. John Tukey agreed to attend the conference as guru du jour. Those of us in attendance were thrilled, and hoped to learn a great deal about multivariate analysis[18] from one of the most important mathematicians and statisticians in the world.

To our surprise, Tukey wasn't much interested in multivariate analysis and mathematical virtuosity. He was interested in finding ways to see more clearly. He did not care much for statistical inference, for testing hypotheses, or for grand mathematical systems. He wanted to make data more easily visible, to make observations clearer, to make it easier to explore. His objection to the correlation coefficient was that it obscured relationships rather than clarifying them. Although he was one of the most capable

---

[18] Mathematical analysis of many variables at once.

mathematicians of his time, much of what he said and did was simple.[19]

Although he had already joined the choir invisible at the time of the Harvard study, Tukey could have pointed out what was wrong in a moment. The Harvard scientists were intent on proving -- or disproving -- their hypothesis to the extent that they obscured looking at their data to see what they revealed, a process which Tukey abhorred. In order to set up a clear test against the null hypothesis, the Harvard group broke their data into two parts -- those who ate the most fat, and those who ate the least. They then computed the occurrences of cancer, heart disease, and the like for the two groups and calculated the differences. These differences turned out not to be statistically significant.

But Americans eat one of the fattiest diets on

---

[19] At the time, I was trying to work out the mathematics of comparing two or more multidimensional non-Euclidean spaces. Over dinner, I described my procedure -- which involved rotating the spaces to a least squares best fit -- to Tukey. He said, "Why do you put all the stress between?" I didn't understand, so he moved closer and repeated, more slowly, "Why do you put all the stress between?" I still had no idea what he was talking about, so he moved close enough for our noses and foreheads to touch, and repeated, very, very slowly, "Why do you put all the stress between?" It took a few days before I realized what he meant -- he meant that I might consider adjusting the measured values in the data to make a better fit, as many psychometricians do. I disagreed with that, and still do.

planet earth, so dividing their subject into two halves produced not a low fat and high fat group, but a high fat group and a very high fat group -- the average amount of fat consumed in their "low fat" group was very high by world standards. Both groups, on the average, consumed enough fat to cause serious health problems.

Even though the average amount of fat consumed by the Harvard sample was quite high, the sample size was so large that there were sure to be many people in the sample who consumed very low amounts of fat. How healthy were they? There's no way to tell, because the Harvard group averaged them altogether by lumping them into two categories. If Tukey had been alive and consulted, I'm sure he would have recommended they forget about the significance test and look more carefully at their data. If they had, they might have seen something like this:

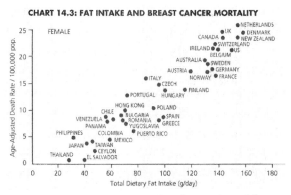

Figure 5: Breast cancer mortality by national fat consumption[3]

This chart is not from the Harvard data, but taken from *The China Study[3]*, and shows the age-adjusted rate of breast cancer mortality per country by the percentage of fat in daily caloric intake. Several things, hidden by the way the Harvard group analyzed their data, are quite obvious here: first, as the percentage of fat in the diet increases, the rate of death from breast cancer increases accordingly. And second, the US has one of the highest rates of fat intake and one of the highest rates of mortality from breast cancer in the world.[20]

The impulse to categorize runs deep in social science, and the decision to divide their sample into two categories -- high fat and low fat -- doomed the Harvard study.

1. P. Cairns, presented at the BCS-HCI '07 Proceedings of the 21st British CHI Group Annual Conference on HCI 2007: People and Computers XXI ...but not as we know it - 2007 (unpublished).

2. R. P. Feynman, *Surely You're Joking, Mr. Feynman.* (Norton, 1997).

3. T. C. Campbell and T. M. Campbell, *The China Study.* (BenBella Books, 2006).

4. L. Dinauer, Doctoral dissertation, University of Maryland, 2003.

---

[20] Campbell, the author of this study, himself considers this correlation to be spurious. Fat consumption is highly correlated with animal protein consumption, which he believes is the real culprit.

Chapter 15: The Rise and Fall of Social Psychology

And now, World War II brings together social scientists from all disciplines to work together. They find the collaboration exciting, and vow to build interdisciplinary programs after the war. Great advances are made, but, in the end, the social psychology movement fails.

Aristotle's psychological model held that the actions of individual men sprang from intentions that were inherent in their substantial form, and harked back in an unbroken chain to the uncaused cause. But, while Aristotle's psychology offers a philosophical theory about why men act one way or another, it does not provide any basis for understanding why one person acts *differently* from another. Following from this fundamental premise, philosophers like Thomas Hobbes assumed that all men would be in competition for the *same goals*, and the only way to avoid universal warfare was for everyone to surrender all power to a "Leviathan," a central government that monopolized force and imposed social order on all.

This Aristotelian archetype resurfaces in 1967 in Blau and Duncan's influential volume, *The American Occupational Structure.*[1] Although clothed in modern language, Blau and Duncan's underlying theory is

Aristotelian: all men (the Blau and Duncan sample is all male) have the same goals which are part of their human nature, and these include the desire to be as successful as possible. The defining factor in how successful they become is the resources they are able to bring to the quest, and these are mainly defined by the status of their father.

Among those who challenged this view was William H. Sewell (1909-2001), a lifelong friend and colleague of Duncan and his early mentor.[21] Sewell had early developed an interest in what he preferred to call "social structure and personality," but what others called "psychological sociology" or "social psychology." He describes his own early education in social psychology as inadequate, but he had read most of Dewey and Cooley and some of Mead, although *Mind, Self and Society* had not been published at the time.[3] These "symbolic interactionists" as they would later be called, offered a revolutionary picture of human behavior derived ultimately from the work of Wilhelm

---

[21] Speaking of individual differences, Sewell and Duncan were quite different: Sewell was proud that he had only missed three national conventions of the American Sociological Association in his long career, while Duncan only attended once, the year Sewell was president. Yet both men are men, with the same substantial form, informed with intentions that go back to Aristotle's uncaused cause. What accounts for the *difference* in their behavior? 2. W. H. Sewell, (University of Wisconsin-Madison Oral History Project, 1976-1988).

Wundt, the first person to describe himself as a psychologist, and one of the last of the chain of Ionian scientists in the newly defined social sciences.

Wundt understood that each individual person lived in a larger cultural context defined by language, customs, religion, family, class, clan, group and other collective phenomena, without which the individual could not be completely understood. Wundt called the study of these phenomena *Völkerpsychologie*, which doesn't translate well into English due to the broad scope of the German word *Volk* and the very different meaning of the English word *folk*. Greenwood[4] favors the translation *social psychology*, even though Wundt himself objected to that translation because the German term *sozialpsychologie* was too close to the then current state of German sociology, which focused on current events rather than the abstract, timeless science Wundt had in mind. Greenwood holds to his position, however, because the current sense of American social psychology is closer to Wundt's original intention than what sozialpsychologie represented in Wundt's time.

One of Wundt's key psychological concepts is the *representation*, which is the agglomeration of sense experiences into concepts. Völkerpsychologie focuses its attention on *collective representations*, which are concepts that transcend individual minds and are the properties of collectives such as religions, cultures, families, classes, clans and the like. Wundt's model implies that the representations of individuals would be strongly influenced by their memberships in different families, clans, religions, and other groups and associations. Mainstream sociology in the twentieth century would say that persons' positions in the social structure would substantially influence their values, norms, beliefs and attitudes, and these, in turn, would influence their behaviors. Ralph Linton would later say that a person's *status* (Latin for *place*) in the social structure would determine his or her role, which each person would be required to act out. Many twenty first century sociologists might replace the term *social structure* with *social network* for a similar but more precise meaning.

Two of Wundt's students played a key role in the development of the social psychology of Sewell's day: Emile Durkheim (1858-1917), who was in Germany from late 1885 to late 1887, with much of that time spent in Leipzig with Wundt, and George Herbert Mead (1863-1931), who spent three years in Germany immersed in Wundt's work. Durkheim's influence extended over all sociologists, including Sewell and Haller, since he is widely considered one of the three "founding fathers" of sociology, along with Karl Marx and Max Weber.

Durkheim adopted Wundt's notion of the collective representation, and made it the cornerstone of his own theory. Following Wundt, he agreed that, in a largely subliminal process, these representations became dissociated from their (sensory) substrate and took on a life of their own. They had a compulsive power over individuals, driving them to higher or lower rates of suicide, or marriage, for example.[5, 6] No one would call Durkheim a social psychologist, but he established a principle essential to any meaningful social psychology, namely that the society and the culture had to pre-exist the individual, and had determining power over the cognitive development of individuals. Thus the source of the representations in individuals is not Aristotle's uncaused cause, nor the substantial form of humans, but rather the collective representations.

Mead provides perhaps the best place to show the significance of the change the concept of Völkerpsychologie, collective representations, social psychology, psychological sociology, symbolic interactionism, or whatever name is used, represents compared to the Aristotelian entelechy. Mead takes over Wundt's notion of symbols as representations, and believes individuals develop their concepts through a symbolic conversation with others. From the first moments of life, other persons converse with the child, point out aspects of experience, and attach symbolic labels to them. Later, as the infant's system of concepts develops, it is able to reflexively examine its own relationship to objects in the world. These objects are defined by symbols already existing in the culture before the child's birth. In an ongoing process, the child begins to develop concepts of "objects", which can be anything in experience that can be "designated or referred to",[7] from something as concrete as a dog, or as effervescent as a momentary feeling, or even nonexistent, such as a griffin. Among the objects of experience is a very special object, one which is always present in every situation -- the individual him or herself.[22] Here Mead departs from Wundt, and presents his most important insight:

> The difficulty is that Wundt presupposes selves as antecedent to the social process in order to explain communication within that process, whereas, on the contrary, selves must be accounted for in terms of the social process, and in terms of communication; and individuals must be brought into essential relation within that process before communication, or the contact between the minds of different individuals, becomes possible. The body is not a self, as such; it becomes a self only when it has developed a mind within the context of social experience. It does not occur to Wundt to account for the existence and development of selves and minds within, or in terms of, the social process of experience; and his presupposition of them as making possible that process, and communication within it, invalidates his analysis of that process. For if, as Wundt does, you presuppose the existence of mind at the start, as explaining or making possible the social process of experience, then the origin of minds and the interaction among minds become mysteries. But if, on the other hand, you regard the social process of experience as prior (in a rudimentary form) to the existence of mind and explain the

---

[22] Wherever you go, there you are.

origin of minds in terms of the interaction among individuals within that process, then not only the origin of minds, but also the interaction among minds (which is thus seen to be internal to their very nature and presupposed by their existence or development at all) cease to seem mysterious or miraculous. Mind arises through communication by a conversation of gestures in a social process or context of experience-not communication through mind.[8]

Here we see an extraordinary departure from Aristotle's notion that every intention of the human mind is contained in its substantial form and ultimately, through an unbroken chain of causality, to the uncaused cause. Mead here argues that society is prior to mind, and that the individual mind arises out of social interaction.

Every object, including the self, is defined in relation to every other object. Among the objects of experience are *behaviors*. Following Wundt, Mead does not consider behavior to be discrete jumps from one state to another state, but rather *ongoing*, a term Mead uses frequently. For Mead, behavior is not emitted in discrete quanta, like carrying buckets of water from place to place, but is a continuous stream, like a river or a garden hose, which can be deflected here and there, but never stops. For the most part, it is not motivated, but subliminal, again following Wundt. Only when the ongoing behavior meets a barrier that prevents it from going on does thinking occur, and then it consists of a symbolic trial and error process in which various methods of overcoming the barrier are "tried out" symbolically. Animals, Mead assumes, cannot do this, but must actually carry out the attempted solutions, and in this, as with his contemporaries, he grossly overestimates the difference in cognitive ability between humans and other animal species. The Athenian identification of Man as the Crown of Creation is still alive even among the pragmatists.

Mead's model anticipates the latter half of the twentieth century's discoveries about neural networks in its assumption that every "object" is defined in terms of its relationship to all other objects. Moreover, it suggests, consistent with the work of W. I. Thomas and Florian Znaniecki, that these relationships vary across social situations, (a tiger in a cage is a different social object from a tiger in your yard). And, as we've seen, behaviors are social objects, and the self is a social object. These, as well, are defined within a situational context.

Behaviors that one performs are behaviors that are consistent with one's definition of oneself in specific situations. The ongoing flux of behavior is deflected as the situation changes, with different social objects entering and leaving the situation. An adolescent may joke with his or her friends, but must stop when an adult is present; students can laugh and play when the teacher is out of the classroom, but must stop when the teacher returns.

Early in life, when small children are engaged in play with other small children, they can see that the other children are selves like themselves; they cry when toys are taken from them, and smile when things work out well for them. The single, individual behaviors of the other children can be seen as models of behaviors for the child himself or herself. The small child can "take the attitude of the other" and understand that it makes one cry when the ball is taken away. In this way, attitudes and behaviors can be taken on one at a time.

But Mead's model can incorporate more than single behaviors. Later in life, when children engage in more elaborate games, learning rates are multiplied. When a child learns to play a complex game like baseball, for example, it is not possible to understand how to play any position in the game without understanding at the same time how to play all the other positions. When a ball is hit to a shortstop, she must understand that the batter will try to run to first base before the ball can be thrown to the first baseman. She must know that the first baseman will move toward first base to receive a throw from the shortstop, while the pitcher will move behind first base to back up the throw. She must know that the situation is completely different if a person is on first base, in which case the most desirable action for the shortstop is to throw as quickly as possible to second base. Unless, of course, there's a runner on second base, in which case the shortstop must try to hold the runner on second, and so on.

The point of all this is that the player of any position cannot understand how to respond to any possible happening unless he or she *understands the role of all the other positions.* This is what Mead means by "taking the role of the other." In this way, the organization of the culture mirrors itself in the self concept of each of its individual members.

Contrast this view with the view of Thomas Hobbes with whom we began this chapter. For Hobbes, the individual is the starting point. Back in the misted past (which none of us can recall) each person must have been at war with every other person, because they all needed and wanted the same things, due to their inherent nature. But how did order come about? This is known as *the problem of social order,* and many important Enlightenment thinkers proposed solutions. Hobbes' own solution, of course, was that, at some point in that unremembered past, humanity must have come to an agreement to surrender all force into the hands of the Leviathan, the all-powerful state. With its monopoly on force, the state could enforce social order.

Mead's far more sophisticated solution is no solution at all, but a denial of the problem. For Mead, individual people are made by society, and, before society, there could be no people. Society itself must have evolved from lower animals, and only as it did was it possible to make those animals into people.[23] People fit into the social order because their beliefs and attitudes -- their self concepts -- are microcosms of the larger society which generated them.

---

[23] This is another expression of Ernst Mach's understanding of how the sensory mechanisms evolved through natural selection.

Mead, of course, was a philosopher, not a scientist, although he had a deep respect for science. His followers were seriously divided on the prospects for a science based on Mead's thoughts. Some, like Herbert Blumer, were strongly opposed to any scientific implementation of Mead's work. Blumer believed that natural language (like English) was much more subtle and precise than any possible mathematical formulation, particularly since human thought and behavior was "too volatile and evanescent" to be described mathematically. Other students, like Manford Kuhn, thought that measurement was possible and desirable, but the simple lists which they interpreted as measurements made it clear they had no idea how to implement measurement of the main concepts.[24]

The split between Blumer and Kuhn was symptomatic of a larger gulf that was opening between the "quantitative" and "qualitative" networks of social scientists. The ensuing debate between these two groups, which continues today, has produced remarkably little light, and even less clash, since both groups express their views only in their own networks, and seldom speak to each other.

The main reason so little progress has been made in resolving the dispute is that neither group has a solid understanding of what science is. For the most part,

---

[24] Kuhn's major attempt at measuring the self concept was the Twenty Statement Test (TST), in which people were asked to write twenty statements about themselves; the resulting list was considered a measurement of the self concept.

qualitative social scientists have no training in science whatsoever, and are at a serious disadvantage when they discuss what science can and cannot accomplish. The model of science that is presented to them is the quasi-ordinal measurement, test against the null hypothesis model. When they argue that this is too blunt a tool to discover anything useful about cognitive and cultural processes, they are right. But when they mistake it for science, they are wrong. Quantitative social scientists generally know no more about science than their qualitative counterparts, since the model they've been taught is what Feynman would call "cargo cult" science -- a model which mimics the form of science, but doesn't get anywhere.[9]

Meanwhile, when World War II broke out, Sewell was called to service as a reserve officer in the US navy, where he worked with Samuel Stouffer and his staff, which was highly interdisciplinary.[25] He also worked with Rensis Likert and his group in the Program Surveys Division of the US Department of Agriculture. After the surrender of Germany, Likert asked Sewell to join an interdisciplinary group that was planning a survey of the effect of strategic bombing on the morale of Japanese civilians. Immediately after the Japanese surrender, Sewell, Likert and his team of psychologists, sociologists, anthropologists, political scientists, sampling experts, and a psychiatrist arrived in Tokyo.

---

[25] The following account is taken from Sewell's own account in Social Psychology Quarterly.3. W. H. Sewell, Social Psychology Quarterly **52** (Jun, 1989), 11 (1989).

Within a year of their arrival in Japan, the team had completed a survey of 3000 Japanese civilians, analyzed the data and presented a final report, *US Strategic Bombing Survey 1947*.

Sewell and his colleagues were impressed by the benefits of their interdisciplinary social psychological approach, and committed themselves to formalizing their relationship into a discipline called variously *social psychology*, *psychological sociology*, or Sewell's preferred term, *social structure and personality*. Programs were started at Michigan, Cornell, Harvard, Berkeley, Yale, Columbia, Minnesota and Wisconsin and other universities. A social psychology program was started at Michigan State University but was called the Department of Communication. The National Opinion Research Center was moved to the University of Chicago, with a broadly expanded program under the direction of Clyde Hart; a new national research center, the Institute for Social Research, was established at the University of Michigan under the direction of Rensis Likert; and the Bureau of Applied Social Research was established at Columbia under the direction of Paul Lazarsfeld. Survey research centers were started at Harvard, Yale, Princeton, Berkeley, Illinois, Minnesota, and Wisconsin, among others. Social psychology support programs were started at the National Institute of Health and several private funding agencies, including Ford, Rockefeller, Carnegie and Sage.

Ten years after the publication of the US Strategic Bombing Survey, a University of Wisconsin Education professor, J. Kenneth Little, conducted a statewide survey of Wisconsin high school seniors about their post high school plans.[10] Sewell was able to obtain the 30,000 questionnaires several years later, and drew a random sample of 10,000 of them. Sewell had maintained an interest in stratification, having devised a scale to measure the socioeconomic status of rural households based on items found in the living room, and he had carried out important research on socialization, having found that early childhood socialization practices were unrelated to personality, contrary to what Freud's theory suggested. Little's data had measures of educational aspirations (intention to go to college), along with demographic information and measures of the seniors' perception of teachers, parents and friends expectations for them. Sewell would resample these students later in life, producing a valuable longitudinal record of their career trajectories.

While this lengthy research program was unfolding, however, the promise of social psychology was waning. In Sewell's own words:

> For a decade great progress was made, particularly in the interdisciplinary training programs at Michigan under the leadership of Theodore Newcomb, with a faculty including Angus Campbell, Dorwin Cartright, J.R.P. French, William Gamson, Daniel Katz, Robert Kahn, Herbert Kellman, Helen Peak, Albert Reiss, Guy E. Swanson, and Howard Schuman, and at Harvard in the new Department of Social Relations headed by Talcott Parsons, with a

faculty including Gordon Allport, R. Freed Bales, George Homans, Alex Inkeles, Clyde Kluckholn and Florence Kluckholn, Gardner Lindzey, Frederic Mosteller, Richard Solomon, and Samuel Stouffer. The growth was less spectacular in other universities, but by no means insignificant. With such a good start, why should the interdisciplinary programs in social psychology have nearly vanished by the late 1960s without ever becoming established in the institutional structure of American universities?[3]

Why indeed? Sewell thought that there were three basic reasons: *first*, the social sciences occupied an inferior position in the social structure of colleges and universities; *second*, the social sciences were in an inferior position in the institutional structure for funding science in the US, and *third*, there were problems in the condition of theory and methodology of social psychology.

The inferior position of the social sciences in university hierarchies and endemic lack of funding are evident, but most important is Sewell's assessment of social psychological and social science theory and method. He concludes that advances in theory and method were too modest to justify the continuous growth of social psychology:

In speculating about the fate of interdisciplinary social psychology, I must point out that no powerful theoretical breakthroughs took place during this period (or for that matter, since then), which might have served as a stimulus to exciting new theoretical developments or new research areas. Advances were made in social psychological theory, but they were modest; although some codification took place, nothing approaching a unified body of social psychological theory emerged. Rather there were improvements in somewhat isolated bodies of special social psychological theories.... There was little or no consolidation of these special theories into a general conceptual scheme for social psychology. The fate of social psychology in this regard was not different from that of the social sciences generally. In fact, it could be argued that none of the social sciences made spectacular progress in developing general theory during these years.[3]

His assessment of advances in method was equally dismal:

Although important improvements were made in research methods during this period, their main effect was to improve the accuracy of our observations rather than to extend our powers of observations. The new computers and computer programs did help us to sort out some of the complexities of social psychological behavior that would have been almost impossible to analyze with earlier techniques.

None of this, however, was sufficient to bring about major theoretical breakthroughs that would fuel great advances in social psychology.[3]

In fact, Sewell's assessment of social science in general as failing to make substantial progress is important, since the decline of social psychology in particular is remarkably parallel to a general decline in the social sciences, as Figure 6 shows:

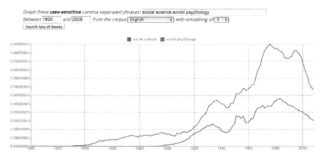

Figure 6: Number of mentions of social science and social psychology in books published between 1800 and 2008

In fact, the decline of social psychology is hardly noticeable when compared to the decline of social sciences like psychology and sociology, as Figure 7 shows:

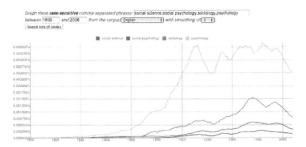

Figure 7: Number of mentions of social science, social psychology, sociology and psychology in books published between 1800 and 2008

To be sure, Sewell, like so many before him, attributes theoretical and methodological difficulties to the special difficulty of social science: "The nature of social psychological phenomena makes such developments very difficult,"[3] but the extraordinary success of science makes it harder and harder to claim that what the social sciences do is difficult by comparison.

The problem is not the subject matter. The problem of the social sciences is their underlying philosophy and the resultant methodology. It looks like science, but it isn't. As Richard Feynman puts it:

> Because of the success of science there is a kind of a, I think a kind of pseudoscience that, social science is an example of a science that is not a science....Now I might be quite wrong. Maybe they do know all this stuff, but I don't think I'm wrong. You see I have the advantage of having found out how hard it is to get to really know something. How careful you have to be about checking the experiments, how easy it is to make mistakes and fool yourself. I know what it means to know something, and therefore I can't... I see how they get their information and I can't believe that they know it, they haven't done the work necessary, haven't done the checks necessary, haven't done the care necessary. I have a great suspicion that they don't know...I

don't know the world very well but that's what I think.[11]

1.   P. Blau and O. D. Duncan, *The American Occupational Structure*. (The Free Press, New York, 1967).

2.   W. H. Sewell, (University of Wisconsin-Madison Oral History Project, 1976-1988).

3.   W. H. Sewell, Social Psychology Quarterly **52** (Jun, 1989), 11 (1989).

4.   J. D. Greenwood, American Psychological Association **6** (February, 2003), 18 (2003).

5.   E. Durkheim, *Essays on Sociology and Philosophy*. (Harper Torchbooks, New York, 1960).

6.   G. Simpson, *Emile Durkheim: Selections From His Work*. (Crowell, New York, 1963).

7.   J. Woelfel, American Journal of Sociology **72**, 1 (1967).

8.   G. H. Mead, *Mind, Self and Society from the perspective of a Social Behaviorist*. (Chicago, 1934).

9.   R. P. Feynman, *Surely You're Joking, Mr. Feynman*. (Norton, 1997).

10.   J. K. Little, 1958.

11.   R. P. Feynman, transcription from You Tube, 2012, http://www.youtube.com/watch?v=IaO69CF5mbY.

Chapter 16: Rediscovering Ionia

And now, while the Dow Demonstration darkens the Wisconsin campus, sociologists there find that increasingly precise measurements show human behavior depends on social forces beyond built in needs and desires. They begin to question the foundations of social science measurement and analysis.

Meanwhile, 22 years before he published his eulogy on the demise of social psychology, in the Chancellor's office in Bascom Hall, William H. Sewell was having some of his darkest hours. Outside his window he could see the Dow demonstration inexorably unfolding in front of the Commerce Building. A few hundred yards west on Observatory Drive, in the attic of Agriculture Hall, one of his former students was conducting a social psychological experiment.

Archibald O. Haller[26] (1926-) began his studies in Sociology at Hamline University in Minnesota after a guidance counselor advised him that his test scores showed he was best suited for either a personnel administrator or a sociologist. It was the first time he had heard the word "sociology." At Hamline he studied

---

[26] Much of what is written here is due to my first hand acquaintance with Haller, but some detail comes from *A Sociologist.* 1.   A. O. Haller, *A Sociologist.* (RAH Press, Buffalo, 2011).

social psychology, particularly what would later be called "symbolic interaction". The leading developer of symbolic interaction theory was George Herbert Mead, a philosopher whose understanding of the role of symbols in mental activity was based on his studies with Wilhelm Wundt, as we have seen.

Haller also took a course in astronomy, where he took an interest in the work of Johannes Kepler. He wondered if Kepler's laws of planetary motion applied as well to the satellites of planets; his calculations for the satellites of Mars differed from Kepler's predictions by about four minutes. Haller was able to make these calculations, not due to anything he learned in his sociology program, but rather because of his service as an electronics technician in the U. S. Navy, where he worked on classified material at Oahu for the latter part of World War II, and because of a previous year as an engineering student at the University of Arizona and a subsequent year as a laboratory assistant at Minnesota Mining and Manufacturing (3M) where he learned some chemistry.

In his last year at Hamline, he ran across the concept of the Pearson Product Moment Correlation. Because his professor knew very little about statistics, he asked Haller to lecture on the topic. He bought a book on correlation by L. L. Thurstone, and another on statistical analysis by Yule and Kendall -- the library at Hamline didn't have them -- which was quite well known at the time and had considerable influence over social science methods. Haller was fascinated by correlation, and it became an important tool in his early work, although later he understood that it would be better to use path coefficients from a technique developed by Sewell Wright called path analysis, which he learned from Otis Dudley Duncan, later to become a lifelong friend.

In 1950, Haller moved to the University of Minnesota, where he studied stratification with Neal Gross, who was a member of a small research group in social stratification that included William H. Sewell and Otis Dudley Duncan. Duncan had originally wanted to study physics, but Sewell convinced him that sociology offered a better opportunity to contribute to science -- a decision which he later regretted. Haller decided that social stratification, and a sub discipline within it, the status attainment process, which engaged his interest in social psychology, were the areas he wanted to study. Gross left Minnesota to go to Harvard, and, when Sewell offered Haller a research assistantship at Wisconsin, he was eager to accept. Not only was Sewell at Wisconsin, but Otis Dudley Duncan was also there, although he left for Chicago before Haller arrived. Moreover, most of the faculty Haller had found interesting were trained at Wisconsin, and he expected a rigorous program.

Haller wrote his dissertation under Sewell on the relationship of personality variables and social stratification variables. Both Sewell and Haller were influenced by the symbolic interactionist model of George Herbert Mead, and were thus under the second generation influence of Wilhelm Wundt, although perhaps without their knowledge. Both were strongly under the influence of Pearson's correlation coefficient, and both were committed to the inferential statistical model of Pearson, Neyman, Fisher and Popper.

Sewell was influential in debunking Freud's theory by showing that adult personality was independent of early childhood socialization, which made him very unpopular with Freudians. Sociologists were very skeptical about psychiatry in general and Freud in particular, not so much because they thought the theory was false, but rather that it wasn't actually a scientific theory, since it made no unambiguous predictions about events or processes that could be observed -- a notion they shared with Karl Popper as we've seen.

Haller, however, was more theoretically focused, and was interested in developing a social psychological theory of status attainment, which is the process by which young people move from the status of their parents to their own adult status in the society.

Otis Dudley Duncan, along with Peter Blau, had written a book called "The American Occupational Structure" that presented a model in which the status men could attain as adults was shown to be affected by the status of their fathers. But the Blau-Duncan model was based on an overly simplistic Aristotelian theory: it assumed that everyone wanted to attain the highest status possible, but were differentially assisted or hindered by their social and economic resources.

Haller thought that the status attainment process was mediated by social psychological variables, particularly the self concept, which in turn was influenced by the expectations of the individual's significant others. If significant others held high expectations for a child, he or she would be expected to hold high expectations for his or her own attainments, which Haller called *aspirations*. These aspirations, in addition to the Blau-Duncan variables, would influence the child's *status trajectory* as he or she moved through life.

Sewell had data -- originally collected by J. Kenneth Little -- which included crude measures of some of these variables from Wisconsin, and Haller had collected similar but more thorough data in Michigan, and Haller convinced Sewell to coauthor a study with himself and a graduate student, Alejandro Portes, who would later go on to be one of the most prominent sociologists of his generation. Haller had by now become acquainted with path analysis through a paper by Duncan, and was comfortable with path diagrams because they resembled the schematic diagrams he knew as a navy electronics technician.

The model that Haller devised suggested that social structural factors, such as the socioeconomic status of the child's family, would affect the expectations that the child's significant others held for him or her, which in turn would influence the child's own self concept, in this case the child's educational and occupational aspirations, which, in turn, would influence the child's later educational and occupational attainments, controlling for what Haller called "measured mental ability." Although both Sewell and Haller were far more sensitive to precise measurement than most social scientists, the nature of their survey sample -- which they did not design -- only allowed for a crude measure of the influence of significant others: a three item index consisting of the students' opinion as to whether their parents expected them to go to college, whether most of their teachers expected them to go to college, and their estimate of whether most of their friends were planning to go to college.

Sewell, Haller & Portes used a technique they learned from Duncan called path analysis, which was developed by Sewell Wright in 1928, which is a set of simultaneous regression equations. Because all the variables in the study were measured with incommensurate scales (measured mental ability, for example, was measured on a three digit scale, and significant other influence on a three point scale), they used standardized data, but, for its time, it was one of the most sophisticated analyses ever attempted in sociology.

In physical science, the model would be tested by whether or not the values of the variables predicted by the theory matched the observed values. But social science theories, this one included, seldom predict actual values, but only gross relationships, e.g., the higher the significant other expectations, the higher the educational and occupational aspirations. By the standards of the time, the theory would be rejected if any of the path coefficients were in the wrong direction (e.g., had negative signs where positive were predicted or vice versa) or were not statistically significant, and third, if the amount of variance[27] accounted for by the model was very low. Finally, if the variance left unexplained in each of the variables by the other variables in the model were not correlated with each other, this would indicated that nothing systematic was left out and the unexplained variance could be considered random.

By these standards, the theory was supported by the data, but by more stringent standards, left much room for improvement. Haller knew, for example, that the measurement of significant others' expectations was

---

[27] The word "variance" is an unfortunate legacy from Galton and Pearson. We've already seen that the correlation coefficient is the cosine of the angle between two vectors. If we drop a perpendicular from the end of one of the vectors to the other vector, we can see that -- if they are correlated -- we've formed a right triangle. We know that the square of the hypotenuse of a right triangle is the sum of the squares of the other two sides. Galton and Pearson call these squares the *variance*. The square of the hypotenuse (the first vector) is considered the total variance. The square of the distance from the origin of the second vector to the point where the perpendicular intersects it is called the "explained variance", while the square of the side that is perpendicular to the second variable is called the "unexplained variance." It's easy to see that, if the correlation is perfect, the angle between the two vectors will be zero and the perpendicular line will vanish, meaning that all the variance is explained; if the correlation is zero the angle is 90 degrees, and none of the variance is explained. As a result of this unfortunate terminology, very few social scientists know that they are dealing with distances and angles.

very crude. First, the expectations of others were not actually measured, only the students' perceptions of them. Second, the actual significant others for each individual were not identified, only the expectations of parents, teachers and peers in general. Third, the scale was only a composite index of three dichotomous variables, which is the lowest form of measurement possible.

Haller proposed a much more ambitious study in which the actual significant others for adolescent children would be identified by name, and the expectations they held for the students would be measured directly and precisely. He had secured significant funding for the project from the National Institute of Education, and had made arrangements with another sociologist to direct the project, but he became ill, and Haller had to find a replacement. I had studied with Sewell, and he served on my doctoral committee through my master's degree until he resigned from all committees to become Chancellor of the University in 1967. Sewell recommended me for the position, and I became the director of the Significant Other Project. Haller and I had never met -- he was in Brazil at the time, and would return there several times during the life of the project.

Although the overarching structure of the project followed directly from Haller's work with Sewell, Portes and Ohlendorf on the status attainment process, there was no theory to suggest who might be significant others, nor any specific plan for identifying significant others and measuring their expectations. Although I had run a laboratory experiment, I had never worked on a field survey before. Although I had had all the advanced statistics and methods classes that were taught at Wisconsin (Sewell told me at my preliminary hearing that I could claim methods and statistics as a specialty) I knew that my understanding was very rudimentary, particularly since Haller and Sewell's previous research used the most advanced statistical techniques known to sociology.

I did two things. First, I scoured the list of unfunded graduate students in the sociology department, and found Edward L. Fink[28], who studied sociology and mathematics at Columbia. Haller agreed and we hired him as a research assistant. Second, I went to the library and found a mathematics textbook that started with "Chapter 0: What is a number?" and I began to read mathematics and science, which has become a lifelong habit.

I devised a crude theory of significant other influence based on a simple categorical scheme which attempted to measure how much information about education and occupations the adolescents received from others, and we constructed questionnaires which asked adolescents to list the names and addresses of

---

[28] Many years later I was teaching a doctoral seminar at the University of Buffalo one evening, and telling about how Ed rescued me on the significant other project. At that point my cell rang, and the students demanded that I answer it. It was Ed. I put him on speakerphone and told him I'd just explained to the class how he had saved me. He said "I saved you? I thought you saved me!"

people who communicated with them about each of the categories. The questionnaire measured the students' occupational aspirations with Haller's Occupational Aspiration Scale, which measured the students' aspirations short term (when your schooling is over) and long term (when you are thirty years old) aspirations, both realistically (the best job you are sure you can get) and idealistically (the job you would most want to have if you could have any job). Educational aspirations were measured with a crude 7 point scale ranging from some high school (students were already in high school) through post graduate degrees.

Another questionnaire was designed for the significant others identified from the students' instrument. They closely paralleled the students' instrument, but were reworded to ask the significant others' expectations for the student.

Analysis of the data was not straightforward. Because the number of significant others for each student varied, each case had a different number of variables. No one had any experience with such a dataset. Because Sewell and Haller were richly connected to the most prominent social scientists, we were able to consult with the world's foremost experts, including people like Duncan, Edgar Borgatta, and others, but no one had any suggestion as to how to proceed. Since the project was nearing its end and a report had to be filed with the National Institute of Education, we agreed to take the mean of the significant others' expectations as a single variable for the purposes of the report, and seek a better solution afterwards. Helcio Saraiva, one of Haller's students working on another project, wrote a special FORTRAN program to compute the means for us, and we then carried out a path analysis.

Surprisingly, this model[2] worked substantially better than the Sewell, Haller and Portes model. "Better" in this case meant that the correlations between significant others' expectations and the students' aspirations were considerably higher, and the unexplained variance was much smaller. None of us expected this because the averaging of expectations was considered to be a desperate maneuver.

The implications of the averaging process were worked out over the next several years. The model was a communication model, in which a person receives "messages" from other people; his or her attitude tends toward the average of these messages. Each message could be conceived as a force pushing or pulling the focal individual's attitude one way or another, but the mean is the place at which those forces balance.[29]

The implications for attitude change are compelling: the formula predicting a new attitude will

---

[29] In preliminary interviews conducted during the significant other project, a young girl was interviewed who claimed to be uninfluenced by others -- she was an independent thinker. When asked how she knew she was an independent thinker, she said that her parents had told her she was. She also planned a career in dressage, a ballet for horses, which seems like a unique career path until you learn that her mother was a ballerina and her father a horse trainer.

be the formula for a new mean given that additional values are added to an old mean. This means that N, the number of messages in the original mean, is an indicator of the stability of the old attitude. If N is very large, a few new messages will have little effect. The amount of attitude change to be expected would then be a function of the difference between the mean of the incoming messages and the old attitude, and the ratio of the number of incoming messages to the old N.[30]

This, of course, assumes that all messages are created equal, which is not obvious. Several experiments, however, showed that the model worked better than alternative approaches to attitude change. John Saltiel directed a three point in time survey over six months that measured students' aspirations and identified their significant others, then contacted the significant others directly using the Wisconsin Significant Other Battery. Six months later the same questionnaires were administered to the same students and various path models were constructed. The best fitting model showed that the amount of attitude change was inversely related to the amount of information (inertial mass) of the old attitude and was independent of the student's subjective certainty, the significant

---

[30] Much of the work in realizing the implications of the averaging procedure was done by John Saltiel, Curtis Mettlin, Donald Hernandez, Kenneth Southwood and myself at the University of Illinois, and was summarized in an unpublished working paper, *A theory of Linear Force Aggregation*. 3.J. Woelfel and D. Hernandez, (Urbana, 1972).

other's average subjective certainty, the degree to which the students liked and trusted the significant others, and the degree of heterogeneity around the significant other's mean expectation[4].

Curtis Mettlin adapted the Woelfel-Haller model to identify the significant others for individuals' attitudes toward smoking and actual smoking behavior. He was able to account for 66% of the variance in the rate of smoking in his sample, which is very high in comparison to other social science approaches.[5] The following year, Mettlin published another study showing that the same model worked well to explain symptoms of stress.[6]

An international study involving three US and one Canadian university showed that the same model predicted attitudes toward marijuana about as well, explaining 55% of the variance in students' attitudes toward marijuana, and 79% of the variance in self-reported use.[7] Still another Canadian study showed that the same linear force aggregation model accounted for 64% of the variance in the attitudes toward French Canadian Separatism of 412 adult education students in Montreal, and also explained 55% of the variance in their behaviors assisting separatist political candidates, and 37% of the variance in their attendance at separatist rallies.[8]

The belief that human thought and behavior is fundamentally inscrutable is deeply imbedded in Western culture, and the idea that attitude change could be described by such a simple model was hard for other social scientists to accept. Danes, et al. attempted to falsify the theory in a series of careful experiments. They concluded,

> Three mathematical models of communication and belief change were proposed and tested: a proportional change model, a belief certainty model, and an accumulated information model. A quick correlational check of the three models suggested that the accumulated information model was the superior with the belief certainty model being the most inferior of the three. Stronger support for the accumulated information model obtained using a more stringent test: a nonlinear bivariate regression which produced visual "plots" of empirical data that nearly duplicated the visual "plots" produced by the theoretical model. The accumulated information model states that belief change is proportional to the discrepancy between the original belief and the belief communicated in the message, and inversely proportional to the amount of information which the receiver has about the topic at the time the message is received. The belief certainty model was the most inferior of the three indicating that the degree to which a receiver is certain in conviction is unrelated to the communication-belief change relationship.[9]

In a second experiment, Danes, et al., tested the model against behavioral theory, balance theory and congruity theory, and concluded that:

> Belief change is predominantly determined by the discrepancy between message and initial belief. However, beliefs based on a large amount of information are more resistant to change, presumably because the receiver is more likely to attend to internal counterarguments.[10]

The notion of internal counterarguments is consistent with the neural network model we discussed in Chapter 1. Since concepts are modeled as sets of interconnected neurons, activating one or more concepts by means of a persuasive message will result in spreading activation of other concepts related to the concepts implicated in the persuasive message. This should be expected to result in a cascade of messages or arguments, as the individual is "reminded" of the other concepts related to the concepts in the persuasive message.

Although we were unaware of it, the implications of averaging messages was being studied by Norman H. Anderson.[11] This is another example of how an idea arises in the collective consciousness and only later diffuses into individual brains. The exact form of the linear aggregation model is not important, however. By far the most important idea to come from the Significant Other Project was the idea that cognitive structures and processes should be measured comparatively. Once this decision was made, the model was subject to comparisons with observations, and deviations from predicted values force changes in the model. When a categorical model is wrong, it's just wrong, and you are clueless as to how to proceed. But when a comparative model is wrong, it's wrong by a measureable amount, and the theory needs to be modified to reduce that error. Although none of us knew it at the time, observations would compel us along the road to a spatial model of cognitive processes. The characteristics of that space would be determined by observations.

As Feynman said, if the theory doesn't agree with experience, it's wrong. And so it is with the Theory of Linear Force Aggregation. But it is wrong by a much slimmer margin than its predecessors, and the errors serve as a guide to improvement.

1. A. O. Haller, *A Sociologist*. (RAH Press, Buffalo, 2011).
2. J. Woelfel and A. Haller, American Sociological Review **36** (1), 74-87 (1971).
3. J. Woelfel and D. Hernandez, (Urbana, 1972).
4. J. Saltiel and J. Woelfel, Human Communication Research **1** (4), 333-344 (1975).
5. C. Mettlin, Journal of Health and Social Behavior **14** (2), 144-152 (1973).
6. C. Mettlin and J. Woelfel, Journal of Health and Social Behavior **15** (4), 311-319 (1974).

7.   J. Woelfel, D. Hernandez and R. Allen, (Retrieved from Galileoco, Galileo Literature #14. (http://www.galileoco.com/CEtestLit/literature.asp) 1973).

8.   J. Woelfel, J. Woelfel, J. Gillham and T. McPhail, Communication Research **1** (3), 243-263 (1974).

9.   J. E. Danes, J. E. Hunter and J. Woelfel Human Communication Research **4** (3), 243-252 (1978).

10.   J. E. Danes, J. E. Hunter and J. Woelfel, in *Mathematical models of attitude change: Vol. 1. Change in single attitudes and cognitive structure*, edited by J. E. Hunter, J. E. Danes and S. H. Cohen (Academic Press, New York, 1984), pp. 204–216.

11.   N. H. Anderson, *Foundation of Information Integration Theory*. (Academic Press, New York, 1981).

Chapter 17: Galileo

And now, sociologists at the University of Illinois attempt to develop a model of cognitive and cultural processes that is completely comparative. They find that adopting a comparative measurement rule results in a spatial model. None of the analysis methods they've learned have trained them to work with a spatial model.

When graduate students at the University of Illinois sociology department started calling our questionnaires "Galileos," we didn't know that fundamental measurements like distance and time were never named after people -- just another indication of the gulf between social scientists and science. When we found that out, we tried to stop people from calling what we did "Galileo", and tried terms like "metric multidimensional scaling" to distinguish ourselves from the "non-metric multidimensional scaling" people. But our non-Euclidean data weren't really metric, either.

Bob Pruzek, a psychometrician at the University at Albany, helped arrange a joint conference of psychometricians and Galileo researchers at Albany, and Ed Fink led off with an introduction to Galileo theory and method, which he called "metric multidimensional scaling." At lunch, I asked Jim Ramsay, then president of the Psychometric Society, what he thought of Ed's talk, and he said he couldn't understand it. Jim Ramsay is a very smart guy who had just finished a sabbatical studying spline theory at the Sorbonne, and I found it hard to believe there was anything he couldn't understand. He explained that multidimensional scaling was something some of his psychometrician friends did, and what we did bore no relationship to that, and that was the source of the confusion. I asked him if we didn't call it multidimensional scaling, what should we call it? He said "Call it Galileo. Everyone else does." And so we do.

In 1954, Abraham Maslow published a book that tried to show that all people had the same needs, arranged in the same hierarchy, and that these were the motivating factors behind all human behavior.[1] As we've seen, the idea that all people are motivated by internal drives and needs is deeply embedded in Western Aristotelian culture. Another deeply held belief, particularly in an America that was predominantly rural until 1940, was that rural life was wholesome and healthful, while urban life had a corrosive effect on morality. But in 1969, Seymour Martin Lipset and Rheinhold Bendix found that rural students had lower levels of occupational attainment than urban students.[2] Because they assumed that every child would naturally have the same aspirations (as did Blau and Duncan as we've already seen), they assumed this meant that the urban environment was more stimulating and led to higher achievement, while rural life was stultifying and depressed attainment. This reversal of what was a common American belief in the soundness of rural life caused considerable concern, particularly among rural sociologists.

It was quickly discovered, however, that the lower attainments of rural youth were completely explained by the much higher proportion of rural youth who planned to become farmers. Since the occupational status of farming is low, it brought down the average level of attainment of rural youth. Apparently, rural youth weren't stultified after all, they just liked farming.

Although it was not realized at the time, the idea that different people had *different* attitudes and aspirations, and that these could be determined not by the inherent nature of people, but instead by social and cultural factors, was a radical repudiation of the Aristotelian notion that all men's intentions were fixed by their nature, and that the human nature of all people was the same.

Sewell, Haller and Portes' status attainment model, therefore, represented a significant change in social science theory. It was one of the earliest and most ambitious tests of the social psychological model of Wundt, Mead and the pragmatists. This model moved from Aristotle's notion that all men (remember Aristotle did not think women were rational) had the same built-in needs and inclinations, to a model where the attitudes of individuals were strongly influenced by information from the environment, and those attitudes went on to account for the differential behavior of individuals.

The Wisconsin Significant Other Project showed that more precise measures of the same variables made very large improvements in how well the theoretical model fit experience. Later work based on the Linear Force Aggregation model developed at the University of Illinois made substantial gains again in the goodness of fit between theory and observation. The Linear Force Aggregation Theory, with its concept of individuals subjected to flows if information from significant others, media and direct observations all represented as vectors in space in a mathematical model was a good beginning, and even its crude measurement model was a long way from J. Kenneth Little's dichotomous measures, e.g., Do most of your teachers expect you to go to college (yes or no), yielding 6.6 bits of information for Little's one bit. Its fit to observations was much better than competing theories.

But, of course, if it doesn't agree with observations, it's wrong, and there were still real difficulties. While work at Illinois was moving away from categorical thinking and closer to a comparative measurement model, the movement was more of a blind groping than a self-conscious search. Crucial aspects of information flow were still measured with Likert type scales. The media study, for example, measured information about marijuana from others in a categorical way:

> The influence of model-type significant others (X23) was measured by the following questionnaire item: How many of your friends smoke marijuana? The response scale was :(a) none, (b) few, (c) some, (d) many, (e) all or

nearly all.[31]   Definer-type Significant Other Influence was measured in two ways. First, for information specifically transmitted about marijuana, a three item index (X38) was constructed, consisting of an item measuring exposure to friends (how frequently do you talk to your friends?), an item measuring coverage of marijuana (how frequently do your conversations with your friends involve marijuana use?). The response alternatives to the last item were (a) highly opposed to marijuana use; (b) opposed; (c) neither for nor against; (d) in favor of; and (e) highly in favor of marijuana use. This item was scored -2 for the first response, -1 for the second, 0 for neutral, +1 for favorable, and +2 for highly favorable. The first two items were scored from 0 (no exposure or no topic coverage) to +4 (nearly continuous exposure and nearly 100% topic coverage). Thus, the product of the three items is an index varying from -32 (nearly continuous intense

---

[31] Did we assume they would all be too stoned to actually *count* them? No, of course not, because we knew that most of the respondents in our survey didn't smoke marijuana at all. We were simply under the influence of the general cultural belief that all measurements should be expressed in categories, which is what we were taught in graduate school. We suspected, based on the results of the Wisconsin Significant Other Study, that we should measure more precisely, but we didn't know how.

negative) to +32 (nearly continuous intense positive). A zero score would result if (a) there was no reported contact with friends, (b) there was no discussion of marijuana among friends, or (c) coverage was neutral, i.e., neither favorable nor unfavorable.[3]

The self concept variable was also categorical:

The theory proposed here suggests that the attitude ultimately governing behavior consists of the relationship an individual sees to exist between himself and the behavior in question. The formulation judged most satisfactory theoretically in this context was "To what extent do you consider yourself a marijuana user?" followed by the response alternatives a) not at all, (b) very little, (c) somewhat, (d) to a large extent, (e) to a very large extent. [3]

An even better illustration of the crude state of development of theory and method at the time was the measure of the dependent variable, frequency of marijuana use, which, of course, could have been recorded as a simple rate, but the culture of the time was to categorize every variables, even those -- like age, for example -- which could be measured on a comparative numerical scale:

Frequency of marijuana use is self reported on the six item scale: 0 = never, 1= less than once a month, 2 = more than once a month, 3 = about once a week, 4 = more than once a week, and 5 = several times a day. This level of accuracy was assumed sufficient for self-reports.[3]

However clumsy, the determination to improve measurement was a legacy of the Wisconsin Significant Other project. The need for better measurement was not obvious then, and some of the most prominent interactionists, such as Herbert Blumer, were steadfastly against mathematical or quantitative descriptions of symbolic interaction, because they believed that natural language was more subtle and precise than mathematics and quantification. Even today in the social sciences the idea that precise measurement is neither possible nor desirable dominates.

And there were other difficulties. The Linear Force Aggregation Model could only work for variables that could be expressed as numbers, such as the number of cigarettes smoked per day, or the number of times one attended a Quebecois rally. But how would one take the average of categorical expectations? If a child's mother expected her child to become a doctor, and her father expected her to become a lawyer, what is the average of doctor and lawyer?

As Jammer[4] shows, humanity's concepts of space developed gradually. What Jammer doesn't show, however, is that they did not diffuse uniformly through society, and those of us in the social sciences had primitive, undeveloped notions of space long after physical scientists had developed their much more advanced understanding.[32] The Illinois group knew that the elementary mathematics of the Linear Force Aggregation theory implied movements in space, but what space?

Much of the time at Illinois was spent learning the basic mathematics and science of space and distance. Charles Osgood, with whom Sewell had worked during World War II, was at Illinois, and he had developed the Semantic Differential, which modeled all beliefs as vectors in a three dimensional space whose principle axes were active-passive, strong-weak, and good-bad. I spoke to him in his office, but he was not helpful in describing how the semantic differential space was constructed, and sent me to see his assistant, Stuart Umpleby. Umpleby was very helpful, but not too keen on the semantic differential space. He said that

---

[32] While we were developing the Linear Force Aggregation model and later the Galileo model, a very common criticism of our colleagues in the social sciences was that a Newtonian model was passé, because we had moved beyond that a long time ago. But it was clear to me that social scientists were still a long way from understanding Newton, and, while contemporary physicists had clearly gone beyond him, we social scientists in general, and I in particular, certainly hadn't. 5. J. Woelfel, in *Communication theory: Eastern and western perspectives* edited by D. L. Kincaid (San Diego: Academic Press, 1987), pp. 299-314.

Osgood had an *a priori* idea of how the space was supposed to look and it took considerable kludging (such as lists of forbidden words that were known not to work, and the like, along with ignoring "defective" cases that didn't work out as Osgood expected) in order to make it (almost) work.[33]

Much more helpful were the technicians at SOUPAC (Statistically Oriented Users Package), a group indigenous to the University of Illinois at Urbana-Champaign who wrote and maintained statistical software to serve the Illinois community. They directed us to factor analysis and multidimensional scaling. Although I had taken a course in factor analysis from Edgar Borgatta at Wisconsin and had (attempted) to work out two factor analyses by hand, I had no idea factor analysis had anything to do with space -- which I believe is true for most social scientists, even those who use factor analysis in their work.

The multidimensional scaling link quickly led us to psychometricians, particularly Warren S. Torgerson,[7] and he in turn led to Young and Householder, who had worked out a correct solution to our problem in 1938.[8] Torgerson also discussed Stevens' ratio, interval, ordinal nominal scaling taxonomy, but indicated that physical scientists only considered the first two

---

[33] Later on, when we understood how to turn distances into spatial coordinates, we were able to see for ourselves that the semantic differential procedure didn't work. 6. J. Woelfel and E. L. Fink, *The measurement of communication processes: Galileo theory and method*. (Academic Press, New York, 1980).

legitimate measurements. This was perhaps the first time I became aware of the fact that most of the measurements most social scientists mainly used would not even be considered measurement in the physical sciences.

For the first time, I determined to find a way to measure cognitive and cultural processes using only comparative scales. To me, cognitive processes were those defined by Mead and the interactionists, and cultural processes were those defined by Emile Durkheim, which I had learned at Wisconsin. Mead and the interactionists saw behavior taking place in *situations* which were defined by the *social objects* in them. A social object, as we have seen, could be anything, real or imaginary, concrete or abstract. Each object was seen to be defined in terms of its relationships to all other objects in the situation. Thus the meanings of concepts, the meaning of the relationships between concepts, the meaning of an individual's self and its relationship to these situational objects could be viewed as objects in space.

In my quest to learn the mathematics needed, I borrowed a book from the Illinois library called *Linear Spaces and Matrices*, and tried to apply what I learned there to the interactionist theory. Much this was recorded later in two working papers, *Procedures for the Precise Measurement of Cultural Processes*,[9] and *Multivariate Analysis as Spatial Representations of Distance*.[10] The first of these began with an indictment of the necessity of representing cognition as a process of categorization, and combined this with the interactionist idea that each object of experience is defined in terms of its relationship to all other objects in a given social situation. In a categorical model, this would be represented by an object-by-object matrix of ones and zeros (as an apple would either be red or not, and so on), but one could just as easily construct a matrix in which each cell represented the difference between two objects on a continuous comparative scale (i.e., how red is it?).

The second half of the paper described the procedures (learned from SOUPAC, Torgerson, and Young and Householder) for projecting these continuous differences onto coordinates. Any physicist would immediately recognize these procedures as a transformation to principle axes -- the eigenvectors of the scalar products of the dissimilarities -- but the long road from the Wisconsin Significant Other Project to the principle axes was a clear indication of the gulf between the social scientist and the physical scientist.

The first test of this model was a simple drawing of Alice in Wonderland.

Drawing of Alice in Wonderland

Undergraduate students were asked to estimate the distances between Alice, the door in the tree, the Cheshire Cat, the March Hare, the Chimney and the Mad Hatter. The matrix of average distances was calculated, and entered into the factor analysis program in SOUPAC. Although there were many options in the way to run the factor analysis -- none of which we understood -- and the options we chose were not optimal, nonetheless the result was plotted on the inside of a large cardboard box, and a hole cut into one of the sides in a spot that approximated the viewpoint of the drawing, and all the objects appeared in pretty much the right places -- ("they agreed pretty nearly").

The next step was to see if the procedures would work with social objects. Several projects were begun simultaneously. One study asked about 1000 undergraduate students (gathered in 7 separate administrations over a six month period) to estimate the dissimilarities among 32 behaviors and the self concept, represented by the term "me" -- a total of 528 paired comparison estimates. Respondents were told that the distance between "sitting" and "lying down" was one "Galileo", then asked how far apart each behavior was from each of the others. Because this questionnaire was so long, it was divided into three parts, and each student responded to one third of the pairs. The results were averaged across students for each pair comparison.

Among the most important things learned from these data was that we did not know how to manage unbounded measurements -- after all, the largest value that can be entered on a five-point Likert type scale is five. No one had heard of Chauvenet's Criterion[34], and none of the extreme values were filtered out, which meant that some values of over 4000 were entered into variables whose mean was about 12 on the average. Another important finding was that undergraduate students could fill out the Galileo paired comparison magnitude estimation scales (remember, social scientists believe that the average person's ability to estimate magnitudes is so poor that they ask people to report numerical data such as age, income, time duration and the like in categories, like age ten or under, 11-19, 20-29, etc., or income less than $10,000 per year; 10,000-19,000, etc., or how long have you lived in Pittsburgh? less than a year, 1-5 years, 6-10 years, etc.).

The 32 behaviors measured in this study were chosen to represent common behaviors that occur very frequently (e.g., smiling, eating, walking, etc.) through to those seldom if ever performed (killing, practicing medicine, marrying). We suspected that seldom performed behaviors should be far from the self, while frequently performed behaviors should be closer. After removing those means obviously distorted by extreme values, this is clearly the case, as Figure 8 shows:

---

[34] Data points whose likelihood of occurrence is less than 1/2n, where n is the number of measurements, are considered suspicious and may be deleted.

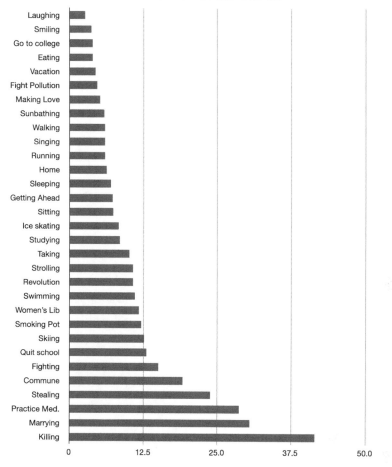

Figure 8: Distances of various behaviors from "me." (Adapted from Wisan[11])

Figure 8 shows clearly that the Galileo, even in its crude first appearance, provides a useful measure of self concept, more useful and predictive than the simple list of 20 categorical responses of the Twenty Statements Test. During the next few years, the exact mathematics of constructing the Galileo space were understood, and several large scale studies established that the precision and reliability of the system exceeded conventional social science measures.[12, 13]

Other real progress was made. Once it was clear that social objects could be precisely and reliably located in a spatial manifold, the solution to the averaging of categorical objects became clear. What is the average of doctor and lawyer? Simple: the average of the coordinates of doctor and lawyer is a point midway between them. The average of Ballerina and Horse Trainer is a point in Galileo space. The average of artist, hairdresser and journalist is the average of their coordinates, which is the midpoint of the triangle connecting them. This model extends the averaging model of the Wisconsin Significant Other project and the Linear Force Aggregation Theory to the discrete case.[14] John Saltiel quickly showed that the theory worked very precisely for discrete occupational choice.[15, 16] In Saltiel's study, when a child's significant others expected specific, discrete occupations for the child, his or her own choice could be accurately predicted to be the occupation nearest the mean of the coordinates of the expectations of his or her significant others.

Saltiel's work takes on special importance because it shows that the categorical model can be subsumed into the comparative model. It is possible to model categorical statements in the comparative Galileo model, but the categorical model cannot express comparative statements. Put in the terms of another cultural split evident in the social sciences, the quantitative Galileo model can express qualitative statements perfectly, but the qualitative model cannot express quantitative statements modeled by Galileo.

This model was quickly generalized by the Michigan State Communication Group into a model which could generate optimal strategies to move any concept from any location to any other location in Galileo space. Since the rate at which behaviors occur is highly inversely correlated with the distance of the behavior from the self-point, moving behaviors closer to the self point increases the rate at which they are performed. Many experiments have examined this idea, with generally positive outcomes.[19-30]

So a theory of cognitive and cultural processes, self-consciously designed to incorporate the attributes of Ionian Science, is supported by preliminary evidence. It appears that a social *science* may indeed be possible.

1. A. Maslow, *Motivation and Personality*. (Harper, News York, 1954).

2. S. M. Lipset and R. Bendix, *Social Mobility in Industrial Society*. (University of California Press, Berkely, 1959).

3. J. Woelfel, D. Hernandez and R. Allen, (Retrieved from Galileoco, Galileo Literature #14. (http://www.galileoco.com/CEtestLit/literature.asp) 1973).

4.   M. Jammer, *Concepts of Space*. (Harvard University Press, Cambridge, 1954).

5.   J. Woelfel, in *Communication theory: Eastern and western perspectives* edited by D. L. Kincaid (San Diego: Academic Press, 1987), pp. 299-314.

6.   J. Woelfel and E. L. Fink, *The measurement of communication processes: Galileo theory and method.* (Academic Press, New York, 1980).

7.   W. Torgerson, *Theory and Method of Scaling.* (Wiley, New York, 1958).

8.   G. Young and A. Householder, Psychometrika **3**, 4 (1938).

9.   J. Woelfel, *Procedure for the precise measurement of cultural processes.* (RAH Press, Amherst, NY, 2009).

10.   J. Woelfel, *Multivariate analysis as spatial representations of distances.* (RAH Press, Amherst, NY, 2009).

11.   G. Wisan, Doctoral dissertation, University of Illinois at Urbana-Champaign, 1972.

12.   J. Gillham and J. Woelfel, Human Communication Research **3** (3), 223-234 (1977).

13.   G. A. Barnett, (Retrieved from Galileoco, Galileo Literature   #18. (http://www.galileoco.com/CEtestLit/literature.asp) 1972).

14.   J. Woelfel, in *Career behavior of special groups* edited by J. S. Picou and R. E. Campbell (Merrill, Columbus, OH, 1975), pp. 41-61.

15.   J. Saltiel, Work and Occupations **15** (3), 334-355 (1988).

16. J. Saltiel, *An application of a social psychological model to the problem of occupational choice.* (RAH Press, Amherst, NY, 2009).

17. R. T. Craig, Doctoral dissertation: Michigan State University, 1976.

18. R. T. Craig, Human Communication Research **3** (4), 309-325 (1977).

19. F. Korzenny, N. J. Stoyanoff, M. Ruiz and A. David, International Journal of Intercultural Relations **4**, 77-95 (1980).

20. S. Chung and E. L. Fink, Human Communication Research **34**, 477-504 (2008).

21. J. Woelfel, E. L. Fink, R. Holmes, M. Cody and J. Taylor, (Retrieved from Galileoco, Galileo Literature #40. (http://www.galileoco.com/CEtestLit/literature. asp) 1974).

22. J. Woelfel, R. A. Holmes, M. J. Cody and E. L. Fink, in *Readings in the Galileo system: Theory, methods, and applications* edited by G. A. Barnett and J. Woelfel (Dubuque, IA: Kendall-Hunt, 1988), pp. 313-332.

23. J. woelfel, C. Meadows and R. Wallace, in *70th Annual Meeting of the Journal of Animal Science* (East Lansing, MI, 1978).

24. K. Rezaei-Moghaddam, E. Karami and J. Woelfel, Journal of Food, Agriculture & Environment **4** (2), 310-319 (2006).

25. Y. Lim, G. A. Barnett and H. Kim, in *International Communication Association conference* (Montreal, Canada, 2008, May).

26. S. Bass, T. Gordon, S. Ruzek and A. Hausman, Biosecurity and Bioterrorism: Biodefense Strategy, Practice, and Science **6** (2), 179-190 (2008).
27. S. Allen, The Forestry Chronicle **81** (1), 381-386 (2005).
28. H. C. DeLeo, MA thesis, Temple University, Philadelphia, PA (1976).
29. G. A. Barnett, in *workshop on metric multidimensional scaling, International Communication Association conference* (Chicago, IL, 1978).
30. G. A. Barnett, K. Serota and J. Taylor, Human Communication Research **2** (3), 227-244 (1976).

Chapter 18: Data Driven Theory

And now, data generated by the comparative Galileo model contradict the most fundamental assumptions of social scientists. They show that the space of cognition is neither Euclidean nor 3 dimensional. Researchers struggle to find ways to reconcile the observations with the assumptions. In the end, they decide to accept the observations and reject the assumptions. They become heretics.

We began this book by noting that, in Social Science, old theories never die -- they don't even fade away. But, not only do old social science theories not die, they don't change. You'll look in vain for a book titled "Recent Developments in Marxism." Max Weber's theory of situses is the same as it was when I learned it in graduate school, and the Theory of Cognitive Dissonance is exactly as it was when Leon Festinger first thought it up. That's because categorical theory is immune from any but the crudest observation. Since that's the case, researchers' attitudes and beliefs are not heavily influenced by observations, and settle on the mean of the information they receive from others. Which of the innumerable social science theories any given social scientist will hold will depend far more on where he or she studied and with whom he or she interacts than on any data. On the other hand, comparative theory, coupled with comparative measurement, is in constant flux, the difference between theoretical predictions and actual observations always visible. Galileo theory and method have evolved from their crude early statement in a constant response to observations.

Ionian comparative theories are continuously modified by observations. So far in this book we've examined the development of the concepts of space from a crude notion of "place" to a cosmology of touching spheres with no intervening space, a space in which "up" is good and "down" is bad; to the empty, flat, limitless space of Kepler and Newton, to the non-Euclidean space of Einstein and the strange space of the quanta. And we've also seen some examples of the spaces of the social sciences: the space of factor analysis, where the dimensions themselves are thought to have meaning, so that, for example, low coordinate values on the x axis mean less intelligence and high coordinate values on the x axis mean high intelligence. The factor space is also a static space in which motion is forbidden: IQ is thought to be fixed at birth and change does not occur.

We've also seen the space implicit in Thurstone's idea of different levels of attitudes constituting positions in an otherwise undefined space. Motion is possible in this space, since attitude change is considered possible, but little else is known about the space other than a statistical estimate of the distance among the most consistent positions measured on an 11 point scale. We've seen Osgood's three dimensional space, in which the first dimension represents good and bad, the second activity and passivity and the third strength and weakness. We've seen the space of early multidimensional scaling, which was largely rejected by the psychometric community because of its high dimensionality and non-Euclidean character, and the response to it, the two or three-dimensional non-metric space in which distances have no meaning and only ordinal position matters.

Along the way, we've seen that the fundamental structure of these spaces is a function of the measurement procedures which form them. Each transformation -- such as the change from Aristotle's notion of "place" to the atomists' notion of particles in space and Newton's absolute, empty flat space to Einstein's curved non-Euclidean space have been necessitate by observations and brought with them much weeping and gnashing of teeth.

And last, we've seen the Galileo space, a space that is formed by measurement procedures self-consciously designed to be exactly the same as those by which all our physical experiences are measured.[35]

---

[35] It bears repeating that treating two sets of experience differently and finding the resulting observations to be different teaches you nothing. In any worthwhile experiment, it is necessary that the same procedures be used across observations. If the observations then differ, you've learned something. Thus it is imperative that we define measurement in the same way for physical and sociological phenomena if we're to discover any differences between them that aren't simple artifacts of the way we observe them.

None of the early Galileo researchers knew much about multidimensional, non Euclidean spaces, and were learning as fast as they could, but the learning curve is very steep. The training of social scientists, even from the most advanced and quantitatively oriented departments, doesn't include the mathematical or scientific tools required to discuss processes in high dimensional non-Euclidean space, or even the tools required to *learn about* high dimensional non-Euclidean spaces.

Most textbooks on the mathematics of Hilbert spaces, Riemann spaces, tensor analysis and the like were beyond the mathematics of even the most advanced Galileo researchers, and they would have to first study lower level texts to learn enough to read them. Even now, the best and brightest graduate students, who are comfortable with statistics and multivariate analyses such as linear regression, path modeling, factor analysis, social network analysis and the like grow apprehensive when they confront tensor analysis and high dimensional Riemann space.

One of the main characteristics of culture is that it constrains the behavior of its members. Although there is a profound sense in which culture gives rise to the concepts which are the vocabulary of thought, much of the constraint is more mundane. In our case, none of the mathematical operations we needed to perform – even once we learned to understand them – could be performed by existing software available to the social scientist. Social scientist are strongly constrained to apply the Pearson, Popper, Stevens tests against the null hypothesis model by the software packages – like SPSS and SAS -- that are available to them. Galileo researchers had to develop special software, a process that occupied over forty years and goes on today. As we learned new mathematical procedures, we then had to learn how to implement them in code.

What we knew about space as it was conceived in the social sciences came from two sources: factor analysis and multidimensional scaling (MDS). As we've seen, the whole point of factor analysis was to identify a *small* set of dimensions that would represent all the factors that determined intelligence (or whatever other topic was being investigated). We also saw that Spearman insisted there was one large dimension, while Thurstone believed there were seven. Charles Osgood was convinced there were three. As far as Galileo researchers were concerned, MDS began with Young and Householder's classic 1938 paper.[1] But even this paper assumed from the outset that a main goal of analysis was to find a solution of *reduced* dimensionality, that is, a space that had fewer dimensions than the raw data, as the abstract from the paper states:

> Necessary and sufficient conditions are given for a set of numbers to be the mutual distances of a set of real points in Euclidean space, and matrices are found whose ranks determine the dimension of the smallest Euclidean space containing such points. Methods are indicated for determining the configuration of these

points, and for approximating to them by points in a space of lower dimensionality.

The idea that there existed a relatively small set of dimensions that could represent all the relationships among a much larger set of variables was the founding assumption of factor analysis and MDS, and a deeply held cultural belief. As we learned how to construct MDS spaces from the psychometricians, we were led to assume by the sheer unanimity of their opinion that this must be the case, so early Galileo software did not even extract all the dimensions possible. Such is the power of culture that it is impossible to be unaffected by this near-unanimous agreement, and early Galileo researchers were alarmed when the data indicated the neighborhoods of the space they studied were high dimensional and non-Euclidean.

In retrospect, what we did seems straightforward. Mead's model suggested that behavior took place in situations, and each situation was defined by the set of social objects which occurred in it. Each of these social objects was defined in terms of its relationship to all the other social objects in the situation. A comparative model means comparison to some (arbitrary) standard. The resulting measurement model followed from these ideas.

Suppose you wanted to measure people's conception of cars.[36] First off, as we learned from

---

[36] Many of the graduate students at the Communication Department at Michigan State University went on to

Bronowski, every part of the universe is (probably) connected to every other part, so we have to "put a box around" the part we want to know about. This corresponds roughly to Mead's notion of the situation. Of course, any social object is, to some extent, defined by its relationship to every other object, but, in practice, we limit our observations to the objects in the situation we are observing. So we might ask our sample of people to list the cars they would actually consider buying within, say, the next 90 days (this is usually called the "consideration set").

Next, we need to know what social objects are in the car-buying situation: what are the other social objects that buyers use to define the cars they choose? We can discover these objects by asking respondents "what's the difference" between pairs of these cars. We are not interested in objects that relate to all the cars in the same way. So we don't need to use concepts like steel, rubber, existing, and the like. We need concepts that differentiate the cars from each other.

As a result of thousands of observations, we know that the frequency with which such objects are mentioned in interviews decreases exponentially, so that, for example, most of the respondents will mention fuel economy, appearance, reliability, cost and so on, but very few will say *spacious glove box*, or the like. Typically, 90% or more of all the mentions will refer to

---

careers in advertising and marketing, and much early research on the Galileo model involved marketing and advertising research in the nearby auto industry.

10 or fifteen concepts. Adding additional objects increases costs for dwindling increases in precision, so we can make a calculated trade off of how much precision is required and how much we are willing or able to pay. Of course, we always include the most important social object of all, the self, usually expressed by the term "yourself."

If we write these concepts (e.g., "good gas mileage, reliable, etc.", the names of the cars and "yourself") down the left side of the page, then write them in the same order along the top of the page, the result is a grid or matrix. In each cell we want a number that tells us how different or "far apart" that pair of concepts are thought to be.[37] (A car that gets good gas mileage should be close to "good gas mileage", an unreliable car should be far from "reliable".) We get these numbers by comparing all these pairs to an arbitrary standard pair, say, good gas mileage and reliable:

Use of the comparative measurement rule where an arbitrary pair of concepts are chosen as a standard and assigned an arbitrary modulus (e.g., good gas mileage and reliable are 100 units apart) then comparing all other possible pairs as ratios to that "criterion pair" (if good gas mileage and reliable are 100 units apart, how far apart are Chevrolet and Volkswagen, Chevrolet and Toyota, and so on) results in

---

[37] In a categorical system, these numbers would all be ones or zeros, indicating that the car had the attribute (1) or did not (0).

a concepts by concepts square grid or matrix of distances or dissimilarities among the set of concepts. These data have the exact form of the distances between cities on a road map.

A common objection to the ratio scaled paired comparison model is that people can't do them reliably and validly. There are several problems with this analysis: first, the notion of validity assumes that there are true, actual distances among the objects, and that the measured values correspond to them. But there are no true, actual distances among these objects -- there are only the opinions of the respondents, and they are filtered through the measurement procedure. Changing the measurement procedure will change the values, but that is true of all observations in science. What we see is nature exposed to our methods of observation. Secondly, by now tens of thousands of observations have shown that the average distances measured this way are quite precise, and in any case much more precise than observations made with typical five point categorical scales. Finally, it is quite wrong to assume, as does the Aristotelian model or Spearman's notion of fixed IQ that respondents can't learn to do this increasingly well. Respondents often volunteer that learning to fill out Galileo questionnaires has sharpened their ability to observe similarities and differences. If we collectively agree that these methods improve individual's and society's collective ability to observe their own cognitive and cultural processes, we will teach them how to do it in school at an early age, as we now teach them to measure time and distance and weight and volume and temperature and a wide variety of "physical" variables that no one can know without instruction.

From the methods we learned from Young and Householder and Torgerson, we are able to calculate the exact space into which all those distances fit perfectly. Remember, as Bronowski and Mead both make clear, albeit in different ways, we are only observing small "neighborhoods" in the overall space and ignoring the rest, but in almost all the neighborhoods we observed, the space is reliably high in dimensions and non-Euclidean.

This is in stark contrast to the way in which theory is constructed in the social sciences. In the comparative model, we are forced to change our theories in the face of contrary data; in the social science model, we change our measured values to fit our preconceived theory. As we saw in Chapter 13, psychometricians assumed *a priori* that the space of cognitive processes must be two or three dimensional and Euclidean, and even developed procedures to distort the measured values to make them fit their preconceived idea of what the space "must" be like.

The process whereby we were led to understand that Galileo space is high dimensional and non-Euclidean tells an interesting story, because it is the story of a group of Athenian social scientists trying to cope in an Ionian comparative mode, which we did not understand very well. Our own concept of space was three dimensional and Euclidean. Of course we were aware (dimly) that Einstein's space was non-Euclidean, but none of us knew precisely what that meant, so it was not a concept we shared with Einstein. People we considered to be the best and brightest among us (i.e., the psychometricians) were telling us that space ought to be two or three dimensional and Euclidean, that our measured values could not be trusted, and even provided clever mathematical computer models for turning our messy high dimensional non-Euclidean observations into easily understood three dimensional Euclidean spaces. This led to differences of opinion in our small research group:

In 1975, Serota, et al. reported on an experiment that showed clear non-Euclidean structure:

> For both groups, six of the 14 characteristic roots (eigenvalues) are negative and large, indicating substantial departures from a linear Euclidean structure. A plausible interpretation for this finding may be the effects of context on the perceptions of concepts. Thus, for example, combining both samples, individuals report the following dissimilarities among the concepts "the rich," "big business," and "me":

"me" and "the rich"	313
"me" and "big business"	237
"big business" and "the rich"	23

No Euclidean triangle can be generated from these figures.[38] Apparently, respondents attend to different aspects of big business and the rich when comparing either to themselves. While this outcome is anticipated by most socio-psychological theory, the perhaps overly-rationalistic views of ideology by major ideological theorists generally fail to consider such discrepancies.[2]

In 1976, Robert Craig published an experiment in which he found:

It can be noted, as an aside, that the fair stability of the imaginary dimensions tends to undermine interpretations of such dimensions as indicating measurement error. Whatever psychological meaning the imaginary dimensions may have,

---

[38] It's easy to prove this. Draw a line 313 units long (any unit will do, as long as 313 of them will fit on your page). Label one end "me" and label the other "the rich". Now take a compass, set it to draw a circle 237 units in diameter (radius of 118.5) and draw a circle whose center is on "me". "Big business" has to be on that circle. Now set your compass for a 23-unit circle (radius of 11.5) and draw another circle with its center on "the rich". "Big business" has to be on that circle, too. You'll see that the circles don't cross, so the triangle won't fit on a flat plane, unless "big business" can be in two places at once.

they are a stable phenomenon, not random
error.[3]

But four years later, Michael Cody published another
experiment, where he disagreed strongly:

> There are two problems with retaining all
> dimensions. First, not all n dimensions are
> reliable...and the inclusion of unreliable
> dimensions is not justifiable. Second, the
> inclusion of "imaginary dimensions" in
> significance testing produces inflated
> correlations...When the loadings of a concept
> increase in the imaginary dimensions, there is an
> increase in "negative distances," which reduces
> the vector length of the concept because the
> negative distances are subtracted from the
> loadings of the concept in the real space. Thus,
> when the cross products are divided by the
> product of vector lengths, the correlations are
> inflated. Also, mathematically treating imaginary
> dimensions as if they were real (Craig, 1977) is a
> poor solution because such a solution increases
> the lengths of all vectors. For these reasons, only
> real and reliable dimensions were used in the
> present research to test the significance of
> hypothesized motions.[4]

But the power of the comparative model lies in
its inexorable ability to push aside prejudice, however
long it might take. Remember the degree of frustration

and desperation with which Max Planck reluctantly modified his equations describing black body radiation to include things he abhorred: a stochastic term from Boltzmann's probabilistic model, and the quantum, later called Planck's Constant. His prejudices were severely violated, but his acceptance of the observations over his prejudices led him to a Nobel Prize. While none of us will be winning a Nobel Prize any time soon, notice the similar angst with which George Barnett and I report the results of our review of the Galileo literature on imaginary eigenvectors and non-Euclidean space:

> The present paper has shown that the Riemannian character of the spatial configurations resulting from these methods cannot be attributed solely to unreliability of measurement. Although several transformations which could eliminate the Riemannian characteristics from these solutions have been discussed, none of them is completely free of problems of its own. While the writers by no means advocate abandoning the investigation of these types of transformations, on the other hand none of them is so compelling as to rule out the simple expedient of dealing with the Riemannian configurations as they are.[5]

The development of Galileo theory and method bears little resemblance to Hempel's idea of hypotheses deduced from nomothetic universal statements.[6] Not only what we know about the structure of Galileo space, but also what we know about movement in the space has always been driven by observations. The first glimpse of any process came in 1970. The first Galileo questionnaire ever made consisted of activities ranging from those performed very frequently, like smiling, walking and the like, then some done less often, like running or fighting, and some that are done very seldom if at all, such as marrying or revolution. We hoped to see whether the most frequently performed behaviors were proportionally closer to the self-point (called "me" in that questionnaire) than those performed less frequently. As we saw, this turned out to be true.

Among the concepts in the questionnaire were some conservation related concepts, and we hoped to see if the location of these concepts might be influenced by the first Earth Day, April 22, 1970. Accordingly, questionnaires were administered before Earth Day, and scheduled to be administered again after. But before the second wave could be put in the field, four students were shot and killed by the Ohio National Guard at Kent State University, President Nixon announced the bombing of Cambodia, and the University of Illinois at Urbana Champaign, like many other universities, went on strike. The Illinois National Guard was called in, and fighting and violence broke out on the campus. Many students and faculty were tear gassed repeatedly, and some were beaten. When the second wave was finally able to be deployed, the most salient discoveries were that fighting had moved very close to "me", while "revolution" had moved far out to the periphery of the space. Clearly, the clash with the

National Guard had given the students a much different notion of "revolution."[7]

Two years later, a study of media effects was interrupted in a similar fashion by unanticipated events when President Nixon announced the bombing of North Vietnam, serious fighting and violence again broke out on the Illinois campus, and George Wallace was shot. Subsequent measurements showed that the War in Vietnam and Crime moved closer to The Most Serious National Problem.[8]

What this meant was that the distance of any behavior from the self point was a good predictor of the likelihood of performing that behavior, and that those distances could change as a result of events. The word "attitude" has been defined in many ways by different theorists, but for clarity of communication among themselves, Galileo researchers define attitude to mean the distance between an object and the self point. Attitude change then is defined as change in the distance between an object and the self point. These events had shown us that attitudes changed in a plausible way as a result of the events. It seemed that, if one could deliberately change attitudes by moving objects toward or away from the self point, it might be possible to alter behavior in a systematic way.

Thanks to John Saltiel's work on occupational choice, we knew that occupations could be regarded as points in a multidimensional space, and that the "average" of two or more occupations was another point in the space whose coordinates were the average of the coordinates of the original occupations.[39] This turns out to be completely general, and the "average" of any set of discrete objects is given by the average of their coordinates. This quite naturally gave rise to the possibility that one might calculate the combined effect of two or more simple messages by averaging their coordinates.[9-12]

Barnett, et. al.[13] quickly arranged to test out this theory by designing a persuasive strategy for a young politician seeking a congressional seat against a well funded incumbent. Although the quantitative data showed the strategy was very successful, perhaps more impressive is the remark by Walter Cronkite when announcing the results of the election on nationwide television: "And, in the 18th congressional district it's Blanchard over Huber...Blanchard over Huber? My, that *is* a surprise!"

Since then, the strategic capabilities of Galileo have been put to use in a wide variety of university[9-11, 14] and private sector research.[15] As a part of its continuing assessment of strategy and tactics in its worldwide mission, particularly in the light of events in Iraq and Afghanistan, the United States Army commissioned a thorough review of all the theories of persuasion and attitude change in the social science literature by its Arroyo Research Center at the Rand Corporation. After a comprehensive review of all the

---

[39] Another way to say this is that each occupation is defined as a position vector in the space, and the average of several occupations is the average of their position vectors.

social science theories of attitude formation and change, the Rand Corporation concluded:

> In many ways, Woelfel's theory was the closest that any social science approach came to providing the basis for an end-to-end engineering solution for planning, conducting, and assessing the impact of communications on attitudes and behaviors. This theory appears to provide a generalized framework for

- visualizing attitude structures in a multidimensional space in which the distance between attitude objects connotes their similarity or dissimilarity, with attitude objects that are judged to be similar closer together and those judged to be dissimilar farther apart
- assessing the degree of similarity in attitude structure within subgroups based on the dispersion around the average positions of attitude objects in multidimensional space
- assessing the level of crystallization, stability, or inertia in attitudes by comparing the average position of attitude objects in space at different time intervals and ascertaining whether differences are accountable to a lack of crystallization in beliefs about the objects or whether they actually reflect the movement of these objects in response to persuasive messages or other factors

- identifying the most effective and efficient campaign themes and messages for changing attitudes in a target audience by identifying where in multidimensional space an attitude object (e.g., "the United States") is relative to other concepts, such as "good" and "evil," and what other attitude objects (e.g., "England") might be associated with "the United States" to move it to a more favorable position.[12]

The iterative process by which Galileo theory and method emerged is a feature of the comparative method. The contrast between comparative Ionian science and categorical Athenian rhetorical model is made clear by Craig's early critique of Galileo theory: Craig published a rhetorical article in which he questioned the utility of the Galileo model. Among the problems he suggested was that the Galileo model was unfalsifiable:

> ...(A)n unfortunate quality of Galileo Theory in its present state is that no experimental test logically can falsify it. Because the theory assumes that laws of motion can be found but does not make any definite claims about the form of those laws, any negative result can be discounted on the grounds that it may have applied the wrong laws or failed to meet the conditions for a test of the right laws.[16]

Here Craig's understanding of theory is that of most social scientists, following Popper: theories are proposed, complete. They must make testable predictions. If these are wrong, the theories are rejected. Through a rigorous testing program, theories that can stand up to the rigorous testing survive in a Darwinian sense. But science does not work that way. The idea of an Ionian theory is some symbolic representation of experience that is connected to experience by a standard, comparative method of observation. When a theory is wrong (as they all are), they are wrong *by a specific amount and in a specific direction*. The theory must then be modified and tested again. Scientific research is an iterative process by which theories become increasingly precise approximations to experience. Occasionally, theory may reach an impasse where modification is not possible, or where the required modifications are cumbersome, as was the case with Ptolemy's model. If a better framework is available, the theory is then replaced by another, but only if a better alternative is available.

This has been the history of the Galileo model. The initial guess that cognitive processes could be represented in a three dimensional Euclidean model was rejected almost immediately. Belief that cognitive processes could be represented in Euclidean space led to many attempts to eliminate observed imaginary components from Galileo solutions, but observations forced researchers to retain them[40]. Now we know that

---

[40] We know now that it is always possible to embed any

we must retain all the dimensions because nature will not conform to our esthetic judgments. All this was forced upon us by observation. The advantage of the Galileo model is that it is connected to experience by a precise comparative measurement model, and any discrepancies between expectations and observations are clearly visible. Not only did no one think at the beginning that Galileo space would be high dimensional and non-Euclidean, or that effective persuasive message strategies could be made by averaging contravariant tensors, but none of us would have known what that meant. The development of the Galileo model was not driven by our hopes or our expectations. It was driven by data.

1. G. Young and A. Householder, Psychometrika **3**, 4 (1938).

2. K. Serota, E. L. Fink, J. Noell and J. Woelfel, in *April 1975 International Communication Association conference* (, Chicago, IL, 1976).

3. R. T. Craig, Human Communication Research **3** (4), 309-325 (1977).

4. M. J. Cody, in *Communicagion Yearbook*, edited by D. Nimmo (Transaction Books, New Brunswick, 1980).

Riemann Space in a Euclidean space of higher dimensionality, but there is no practical advantage to doing so.

5.   J. Woelfel and G. A. Barnett, Quality and Quantity **16**, 469-491 (1982).

6.   J. Fetzer, in *The Stanford Encyclopedia of Philosophy*, edited by E. N. Zalta (Stanford University Press, Stanford, 2012).

7.   G. Wisan, Doctoral dissertation, University of Illinois at Urbana-Champaign, 1972.

8.   J. Woelfel and G. A. Barnett, in *special session on socialization and the media in the Mass Communication Division of the International Communication Association conference* (, New Orleans, LA, 1974, April).

9.   F. Korzenny, N. J. Stoyanoff, M. Ruiz and A. David, International Journal of Intercultural Relations **4**, 77-95 (1980).

10. J. Woelfel, R. A. Holmes, M. J. Cody and E. L. Fink, in *Readings in the Galileo system: Theory, methods, and applications* edited by G. A. Barnett and J. Woelfel (Dubuque, IA: Kendall-Hunt, 1988), pp. 313-332.

11. Y. Lim, University at Buffalo, State University of New York, 2008.

12. E. Larson, R. Darilek, D. Gibran, B. Nichiporuk, A. Richardson, L. Schwartz and C. Thurston, *Foundations of effective influence operations: A framework for enhancing army capabilities.* (RAND Corporation. Available online at http://www.rand.org/pubs/monographs/2009/ RAND_MG654.pdf Santa Monica, CA, 2009).

13. G. A. Barnett, in *workshop on metric multidimensional scaling, International*

Communication   Association   conference   (, Chicago, IL, 1978).

14.   S. Chung and E. L. Fink, Human Communication Research **34**, 477-504 (2008).

15.   J. Woelfel and N. Stoyanoff, in *The role of communication in business transactions and relationships*, edited by M. Hinner (Peter Lang, Berlin, Germany, 2007), Vol. 3, pp. 433-462.

16.   R. T. Craig, Communication Monographs **50** (4), 395-412 (1983).

Chapter 19: Compared to What?

And now, the newfound ability to project cultural beliefs and attitudes onto mathematical coordinates makes it possible to transform views from one frame of reference to another. The comparative model lets us compare our beliefs and attitudes on the same frame of reference, and open up new avenues of understanding.

In Chapter 2, back at the very beginning of our search for the culture of science, Schrödinger found in the Ionian scientists the second stone which -- after the idea that the world can be understood without resorting to magic or spiritual factors -- he believed to be the foundation of the scientific method: the idea that "...the 'understander'" (the subject of cognizance)..." is to be excluded from "...the rational world picture that is to be constructed."[1]

One of the cornerstones of 20th Century social science, however, is that all perceptions are filtered through individual "frames of reference," so that each individual's view of the world is unique. In 1936, Muzafer Sherif showed that individuals' perceptions were filtered through a "frame of reference." In a classic experiment, he placed experimental subjects in a totally dark room, thus depriving them of any visual frame of reference, and projected a dot of white light on a screen in front of them. Due to involuntary movements of the eye muscles, the dot often appears to move. He then showed that he could influence subjects' estimates of how far the dot moved by introducing confederates posing as subjects. If these confederates reported large motions, subjects tended to increase their estimates, while when confederates reported smaller motions, subjects' estimates dropped.[2]

In 1955, Solomon Asch reported a similar experiment, where subjects were asked to judge the relative length of sets of three lines. For each set of three, subjects were asked to judge which line was longest. When confederates posing as subjects gave incorrect responses, the subjects' often agreed with the confederates' incorrect judgment.[3]

Since that time, the idea that people's perceptions are filtered through a "frame of reference" has become a commonplace notion throughout the social sciences. Particularly in the field of Communication, the fact that different individuals' perceptions are influenced by their personal and cultural frames of reference is understood as a formidable barrier to accurate communication. A central thesis of George Herbert Mead's model is that each object has a meaning only in reference of the other objects in a situation, and that the meaning of the "same" object will necessarily differ from one person to the next and from one situation to the next. This inherent "subjectivity" has led many social scientists to abandon efforts toward "objective" science -- science in which the observer is excluded. In a categorical model, when different observers make observations in different frames of reference, their observations are different, but there is no way to establish in what way or by how much the different reference frames influenced the observations.

But this is too strong a position. Objectivity does not need to mean that the observer is indifferent to the observations, or that the observations do not depend on the method of observation. All it need mean is that methods of observation exist which yield the same outcomes no matter who makes them.

No one knows how to make precise transformations from one categorical reference frame to another, but the most fundamental operation of the Galileo model is its ability to transform categories onto continuous, comparative coordinate systems. Transformations across comparative reference frames or coordinate systems have been well understood since Galileo, and, in fact, such transformations are generally called "Galilean transformations." The modifications to Galilean transformations made necessary by the observed constancy of the speed of light have been known for over a hundred years, and are usually called Lorentz or Lorentz-Fitzgerald transformations.

In a comparative model, it is possible to transform across different reference frames to a common reference frame. As social scientists, none of the early developers of the Galileo model had a good understanding that this problem lies at the core of physical science as well, and, over a long time, physicists and mathematicians have developed procedures for transforming observations across reference frames. Although it took us several years to work out a solution, in retrospect it seems straightforward. By projecting the distance estimates of respondents onto a coordinate reference frame, it became possible to utilize the rotations, translations and reflections developed by physical science to transform observations across multiple observers in Galileo space. To be sure, rotations, translations and reflections in high dimensional non-Euclidean space are mathematical operations that don't come in the social scientists' everyday working toolkit, and we blundered about considerably while learning how to do it, but it is now possible to project the observations of different observers from different cultural reference frames onto a standard Galileo reference frame and compare them without the artifactual differences resulting from the different frames of reference.[4-7]

Figure 9 shows the conceptions of different emotions across observers from three different cultures: Korean, Hindi and the United States. Three individual people, one a native Korean speaker, another a native Hindi speaker, and the third a native English speaker, filled out a Galileo questionnaire in their native languages. The resulting distance matrices were then projected onto Galileo coordinates and rotated to least squares best fit on each other.

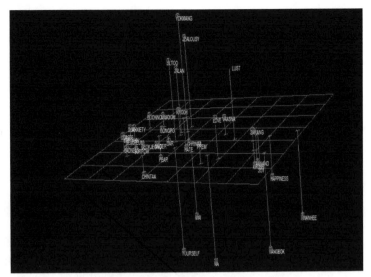

Figure 9: Comparison of emotions across cultures

At the bottom center you can see the Hindi, Korean and English words for "yourself." Although a detailed analysis is beyond the scope of this chapter, transforming multiple individual frames of reference onto the same coordinate reference frame makes it possible to calculate a single number which gives the overall difference between any two patterns, as well as a number for each concept individually.

The Galileo rotation algorithm is not limited to transforming to stationary reference frames. The same transformation makes it possible to maintain a common reference frame across multiple measurement sessions so that time series processes can be represented.[4] In 1989, Galileo researchers published a paper showing how the Galileo rotation algorithm could distinguish the stationary character of the numbers on the face of a clock from the moving ends of the hour, minute and second hand.[6] Three years later, another article showed how the Galileo rotator could solve for cyclical processes, holding the location of the days of the week (Sunday, Monday, etc.) constant while allowing the self point of respondents to move about the closed figure made up of the days.[7]

Although limited in scope and at a very early level of development, the comparative Galileo system is capable of modeling dynamic processes, allowing for the definition of the main variables of dynamic systems: velocity, acceleration, force, mass, energy and work. These variables are not, however, properties of the phenomena under study, but rather properties of the comparative method by which it is studied. The adoption of a comparative measurement rule and transformation of the resulting measurements to a mathematical coordinates system allows us to seek transformations which produce reference frames with desirable properties, such, for example, as reference frames in which the total action is minimized.[8]

1.   E. Schrödinger, *Nature and the Greeks and Science and Humanism.* (Cambridge University Press, Cambridge, 2002).

2.   M. Sherif, *The Psychology of Social Norms.* (Harper and Row, New York, 1936).

3.   S. E. Asch, Scientific American **193**, 5 (1955).

4.   J. Woelfel, R. Holmes and D. Kincaid, (Retrieved from Galileoco, Galileo Literature #19.

(http://www.galileoco.com/CEtestLit/literature.asp)
1980).

5.   J. Woelfel and E. L. Fink, *The measurement of communication processes: Galileo theory and method.* (Academic Press, New York, 1980).

6.   J. Woelfel, G. A. Barnett, R. Pruzek and R. Zimmelman, Quality & Quantity **23**, 3-20 (1989).

7.   J. Woelfel and G. A. Barnett, Quality and Quantity **26**, 367-381 (1992).

8.   J. Woelfel, *Variational Principles of Communication.* (RAH Press, Birmingham, MI, 1988).

Chapter 20: Patterns

And now, as hundreds, then thousands, then tens of thousands of observations accumulate, Galileo scientists begin to see that the collective consciousness can identify, store and retrieve very complicated patterns that are beyond the capacity of single individuals. They are able to observe the process of collective concept formation and change.

The Galileo group of scientists was by no means the first people who recognized the need to break free of the philosophical stranglehold of the Athenians. Alfred Korzybski and his General Semantics was based on a rejection of Aristotelian thinking, and authors like A. E. Van Vogt even wrote books with the phrase "Null A" (signifying non-Aristotelian) in the title. But freeing oneself from a philosophy in which you were born and raised is easier said than done. Gilbert Gosseyn (go sane?), the hero of several of Van Vogt's novels, is allegedly at an advantage because of his null A training, but there isn't much evidence of non-Aristotelian thinking in the books, and Gosseyn's advantage seems to be mainly that he has two brains, which enable him to transport himself instantly across even cosmic distances, and the fact that whenever he dies, he wakes up in a fresh body that has been stored and waiting for him as needed.

Freeing yourself from prejudices built into your way of thinking is not so easy. If you succeed in eliminating the way of thinking you learned without developing a new one, you won't be able to think at all. (I was once asked by a journal to review a book. The author asked me to forget everything I knew so I could understand the book without any preconceptions. I sent in a review that said I forgot I was reviewing the book, but it never ran...)

In classical debate, there are two essential components, the *need* and the *plan.* To win the debate, the affirmative team must show that there is a serious need, that the need is inherent in the status quo, and that they have a feasible plan that can resolve the need without causing greater harm than benefit. To win the debate, the negative team must establish either that there is no need, or that the need is not inherent in the status quo, or that the plan isn't feasible or won't work, or that the plan causes greater harm than it prevents. If all else fails, the negative may present a counter plan, but this is seldom a winning strategy. Establishing the need is usually the easier part of the debate, but constructing a feasible plan that will resolve the need without causing serious harm is never easy. General Semantics had a very slender plan, and A. E. Van Vogt descended into Dianetics before abandoning his quest altogether.

In the Galileo model, the *need* is that the Aristotelian categorical model underlying social science methodology is too blunt to allow meaningful observation, which means that theories can never be precisely compared to observations. The *plan* is the Ionian belief that cognitive and cultural processes can be understood through observations made with the comparative method, followed by communication with other scientists who must check them. As symbolic models of these observations emerge, they need to be checked against observation and altered when they don't conform to observations.

By making these measurements and adhering to them, a model of cognitive structures and processes slowly emerged, but we were not always quick to recognize it. As we've seen, our early prejudice in favor of low dimensional Euclidean space took ten years to wither away in the face of consistently high dimensional non-Euclidean observations.

It took longer to realize that the Galileo system was revealing amazing capacities of the collective consciousness that we hadn't yet recognized. One of the first experiments to show this was done by Robert Craig. He had subjects read a message that said Brazil and Portugal were similar, and he added several reasons -- e.g., they were both Catholic countries, they both speak Portuguese, and so on. He expected to see Brazil and Portugal approach each other along the line segment connecting them. They did grow closer, but not along the line segment connecting them. Craig's sample had recorded all the information in Craig's "noisy" messages, not just that Brazil and Portugal were similar, and could reliably display it a week after reading it once. Although we couldn't measure it precisely because not all of the concepts in the messages were in the questionnaire, there were enough that Craig's post hoc analysis could show some of the effects. It's safe to say that very few, if any, of the individuals in the sample could do that, but the collective could.[1]

At the time, no one took note of the pattern recognition and storage capability of Craig's 64 undergraduate students, who collectively could remember a complex pattern of information a week later after reading it only once. But soon more impressive evidence turned up, again by accident. Scientists at the Communication Department at the State University of New York at Albany and the Communication Institute at the East West Center in Honolulu were interested in the process of learning and forgetting in the collective consciousness. They constructed a complicated random message about six fictional persons. They assigned attributes to these fictional persons by a random process of coin tosses. The resulting message was:

> - Sue is somewhat short. She is somewhat liberal politically. She is a very sloppy dresser. She is very unfriendly. She is very intelligent.
> - Bob is somewhat tall. He is somewhat liberal politically. He is a very sloppy dresser. He is somewhat unfriendly. He is very unintelligent.
> - Mary is very short. She is very liberal politically. She is a somewhat sloppy dresser. She is very unfriendly. She is very unintelligent.

- Dave is very tall. He is very liberal politically. He is a very sloppy dresser. He is somewhat unfriendly. He is very unintelligent.
- Lisa is very short. She is somewhat conservative politically. She is a somewhat neat dresser. She is somewhat intelligent.

In a pretest, 75 undergraduate students at the State University of New York at Albany filled out a Galileo pair comparison questionnaire estimating how different each of these fictional people was from all the others. Each student was asked by a random process either to fill out the questionnaire immediately after reading the message, or to wait one hour, one day, or one week before filling out the questionnaire. The researchers expected to see the space grow larger as the respondents learned to differentiate among the fictional persons, then shrink again as they forgot the differences among them.

What they found was that the space clearly grew from the immediate administration to the one hour later administration, then shrunk somewhat after one day, but, curiously, showed no further shrinkage one week later. The structure of the Galileo space after one week was very similar to the structure of the space after one hour or one day. The collective represented by the respondents clearly remembered the paragraph as well after one week as they did after one day, although it is unlikely that any of the individual respondents could remember the paragraph (the authors did not test the individual respondents' memories).[2]

To get a more precise track of the size and structure of the space over time, a second experiment was designed, using the same message and the same measurement instrument, but this time randomly varying the delay period to one hour intervals beginning one half hour after reading the message. Four hundred and seventy one students participated in this second phase.

While the investigators theorized that the space would expand during learning and contract during forgetting, what they found was that the space remained about the same size until about nine hours after reading the message, then increased *six fold* in size between nine and eleven hours later, then contracted to the original size by 14 hours, and remained about the same size thereafter. Apart from differences in size, the pattern of the array of names remained fairly constant across time, indicating that the collective group remembered the message, and the Galileo was sufficiently precise to detect it.[3]

These results were so unexpected that another independent research group replicated the study one year later. Five hundred and fifty undergraduate students participated in this third study. Again, the results showed a period of latency of about nine hours, with a six fold expansion in length between nine and eleven hours, falling back to initial size by about fourteen hours.[3]

The real significance of these experiments was not fully appreciated at the time. In a typical message generation experiment, respondents would typically receive between one and three assertions, such as "Pigs are beneficial and attractive," and effects were expected to be limited to the concepts in the message. But in these experiments, respondents receive *30 random assertions* -- six random assertions about each of five fictitious people, and observations show that the respondents collectively remember all 30 of them even a week afterwards, a feat that few of even the most accomplished individuals could achieve.

About the time this research was published, Johnson-Laird published a book[4] describing what he called indefinite or indeterminate language, by which he means that readers are led to believe they have a picture of a scene in their minds, but in fact they do not. He gives the following example:

> I have a very small bedroom with a window overlooking the heath. There is a single bed against the wall and opposite it a gas fire with a gas ring for boiling a kettle. The room is so small that I sit on the bed to cook. The only other furniture in the room is a bookcase on one side of the gas fire next to the window -- it's got all my books on it and my portable radio -- and a wardrobe. It stands against the wall just next to the door, which opens almost directly onto the head of my bed.

Although it's easy to show that, after reading this paragraph, the individual reader does not have a clear picture of the room in mind, it is not so for the collective representations of a number of readers. In an experiment, the preceding paragraph was read aloud to 42 undergraduate communication students at the University at Albany, State University of New York. They then filled out a Galileo pair comparison questionnaire listing the objects in the paragraph (e.g., the gas ring, the bed, the bookcase, etc.) and the self point (yourself).

Each student was then directed to draw a picture of the room, and then to compare his or her picture to the pictures drawn by the other 41 students. This caused a lot of nervous laughter, and the students agreed that no two pictures were alike. But the Galileo program produced a picture based on the averaged responses of the students, and this, of course, is a representation of the room. Randomly dividing the data into two halves and rotating them to least squares best fit produces two nearly identical rooms that are not significantly different. Dividing the data by gender produces a room as conceived by the woman and another as conceived by the men. These are very similar except for the location of the self point, with men placing themselves at the door and women placing themselves on the bed.

In another experiment, 64 undergraduate communication students from the same university read a paragraph describing six fictitious pianos, labeled A through F. By a random coin-toss process, each of the six pianos was assigned one of four levels of three characteristics: by a coin toss, a piano had either a rich tone or a thin tone. By another coin toss, its tone was either very rich or rich, or very thin or thin. The same process was used to assign each piano a size (very large, large, small, very small) and action (from very uneven, uneven, even and very even), which yielded 364 possible different pianos.

After hearing the paragraph read aloud, the students filled out a Galileo pair comparison questionnaire which included the names of the pianos (A through F), and all possible attributions, i.e., very rich tone, rich tone, small, very small, etc., and the self point (yourself) for a total of 19 concepts and 171 pair comparisons.

Although this was a very simple experiment, a great deal was learned from it. First, as was the case with Johnson-Laird's room, it showed that even a small collection of individuals (64 in this case) could detect and store the pattern of information in the complicated message even after hearing it only once. Again, randomly split halves revealed no significant differences between the halves. Equally important, it showed that the concept of categorical scaling -- Likert and Likert-type scales -- is not viable. The data showed that the categorical scales -- very large, large, small, very small; very rich, rich, thin, very thin, and very even, even, uneven and very uneven -- are not linear in two senses. First, the four levels of each of the scales do not lie on a straight line in the space, and second, the distances among the four levels are not equal.

While both experiments showed that the collective consciousness is capable of recognizing, storing and retrieving the complicated pattern in the message, the piano experiment shows that there is systematic distortion in the stored pattern: extreme values have a tendency to be truncated. In addition to the Galileo questionnaire, respondents also responded to a test which asked them to check off the levels of the three attributes for each of the six pianos. When a piano was described in a non-extreme way -- e.g., piano A is *large* -- two thirds of the respondents recall it correctly. But when the piano is described as extreme -- e.g., piano B is *very even* -- only one in five recalls it as being *very even*, while about two in five describe it as *even*. This experiment has been replicated many times.

These results challenge the foundations of everyday measurement in the social sciences. They show clearly that the tick marks on a categorical scale -- e.g., strongly agree, agree, neutral, disagree, strongly disagree and the like -- *cannot be assumed to lie on a line, nor can they be assumed to be equidistant.* More important: they tell a great deal about the character of the collective consciousness. When measured by an Ionian comparative rule, it is not a flat, Euclidean space spanned by one-dimensional scales that coincide with the dimensions. Rather, each individual point of any such scale must be located as a point in the space, and the likelihood that any three of them lie on a straight line or are equidistant from their neighbors is vanishingly small.

We've also learned that the collective consciousness can recognize, store and retrieve even complicated patterns very quickly, and retain them over long periods without loss, but that the patterns are systematically altered in the process, particularly by attenuating larger distances. But, at this time, the model is simply a mathematical model. By 1988, however, Galileo researchers were beginning to understand that the substrate of the collective consciousness lay in the collective neural network in human brains linked together by communication.[2]

1. R. T. Craig, Human Communication Research **3** (4), 309-325 (1977).
2. J. Woelfel, S. Danielsen and J. A. Yum, *Cognitive theory of collective consciousness.* (RAH Press, Amherst, NY, 2009).
3. J. Woelfel, B. J. Newton, R. A. Holmes, D. L. Kincaid and J. Lee, Quality and Quantity **20**, 133-145 (1986).
4. P. N. Johnson-Laird, *Mental models: Towards a cognitive science of language, inference, and consciousness.* (Harvard University Press, Cambridge, MA, 1983).

Chapter 21: Spot, Rover, Oresme, Indstar, Catpac and Wölfpak

And now, computer scientists square off along the lines dividing the Athenian purposive model and the Ionian descriptive model. On the Athenian side is Marvin Minsky and the Artificial Intelligence community, who believe all action is goal oriented, and build intelligent machines in that model. On the other, Frank Rosenblatt and the artificial neural network community.

The Department of Rhetoric and Communication (RCO) at the University at Albany, State University of New York, began transitioning to a Department of Communication during the 1980s. One of my first goals of the transition was to introduce a non-technical undergraduate and graduate student body to computing. In the early 1980s, there were no personal computers, and no video terminals. Interaction with the computer was mainly via punched Hollerith cards, or the latest marvel of the time, the DECwriter, a typewriter-like keyboard connected to a 132 column printer.

Email was not yet invented, but the Albany Sperry-Univac computer had a messaging system, and the graduate students' first assignment was to send a message to another student via computer, and hand in the printout from the DECwriter. So much has the world changed since then that few readers will remember Hollerith cards, or DECwriters. But, although the computer has made it possible to make calculations much faster, they are, by and large, the same calculations we made with the Sperry Univac, and neither traditional social science methodology nor traditional social science theory has improved in any fundamental way.

The operating system of the computer was called Exec 8, and it was largely beyond the skills of the first generation students, so a shell called "SPOT" was written (by an undergraduate named Scott Danielsen and myself) to help them familiarize themselves with computing and overcome what was, for some, real fear. Whenever a communication student logged onto the computer, he or she was intercepted by the shell, and all communication with the Sperry Univac was through the shell.

Each user had his or her own "logon folder" which contained personal information about the user. Spot could access that folder, and speak directly to the user: "Hello, Bob, what would you like to do today?" The user could answer in any language, and might say "I'd like to edit a file." At first, of course, SPOT would not know how to do that, so it would say "I don't know how to do that, Bob. Could you show me?" The user (or one of the staff, usually) would then type the actual command, which, in this case, would be @ed *filename*, where *filename* is the name of the file you wish to edit.

SPOT would then store the statement "I'd like to edit a file" in a stack, and parse the statement and store each word in the command in another stack. SPOT would then form a link between the statement "I'd like to edit a file." and the command *@ed filename*, ask "What file would you like to edit?" then execute the command.

The next time the user asked to edit a file, three possibilities might exist: first, the user could write the exact statement "I'd like to edit a file," which SPOT would recognize and immediately ask "What file would you like to edit," and so on. Secondly, the user might write something slightly different, like "I want to edit my file," or "I want to edot my file." SPOT would not recognize this statement, so it would parse the statement and compare the words in the statement with the words parsed out of the original statement. If any of these matched, SPOT would ask "You'd like to edit a file?" If the user answered yes, SPOT would execute the editing sequence. If not, SPOT would check all other words, and, if it came up empty, would cycle back to the "I don't know how to do that, Bob. Could you show me?" routine. Each time a command was executed, it would be moved to the top of the stack, while all others fell one position lower in the stack, so that the most recently issued commands would always be near the top of the stack, and thus be the commands searched first. Commands that were not used in fifty cycles would drop out of the bottom of the stack and be "forgotten."

Over time, SPOT became a very convincing work partner, and most students reacted to it as if it were a person. One typical sequence is the way the source code of the program itself was modified. The user would enter "surgery" and SPOT would say "Oh, No!" and launch the source code editor. When the editor was exited, SPOT would launch the compiler, the program would be compiled and linked, then re-executed. It would then say, "It's a miracle!" and be ready to work.

Two things are interesting about SPOT: first, it was a simulation of George Herbert Mead's self-concept model. SPOT built up a conception of itself through interaction with the Albany communication community. Its self-concept was different for each student who interacted with it, and consisted of sets of behaviors that were activated depending on the appropriateness of the behavior to the situation. SPOT did not perform behaviors that it "wanted" but rather behaviors that were the best fit to the situation, as defined by previous experiences which took the form of symbolic interactions with others.

Secondly, although no one at the time realized it, SPOT was a crude self-organizing neural network which added nodes and links among nodes as it communicated with the community. It was not preprogrammed, like the artificial intelligence systems of its time, but rather developed its self-concept -- its relationship to persons and activities in situations, through symbolic interaction with others. Thus, its self-concept differed for each user with whom it interacted, and grew and changed as interaction proceeded.

SPOT modeled the most important point of contention between Wundt and Mead: Mead understood that Wundt's theory presupposed the existence of functional individual personalities, but did not account for their origin. In this, Wundt's psychology retains the core of Aristotle's philosophy -- the nature of persons is built into them from the first moment of their existence through their substantial form, which in turn is a result of an unbroken chain of causality back to the uncaused cause. Mead, however, insists that the society must precede the individual, and indeed that the individual is developed through social interaction, as was SPOT.

When SPOT was initially deployed, it had no functionality whatever -- no inbuilt goals, drives, attitudes or beliefs. Through the process of symbolic interaction with the communication community at Albany, SPOT built up an expert repertoire of behaviors. It did not, of course, have any capacity for self awareness, and was a very primitive neural network, but the real point is that such a rudimentary neural network could exhibit very convincing human-like characteristics. Many users were quite fond of SPOT, as was I.

None of us knew that SPOT was a neural network, and none of us had heard the term "neural network." SPOT was written to emulate symbolic interaction theory as closely as possible within the limits of our skill and resources. It was not an effort to model the human brain, about which we knew nothing.

But others did, and the relentless progress of Ionian science produced ever more precise and useful observations of the human brain. In 1873, Camillo Golgi invented a revolutionary method of staining brain tissue called "black reaction" staining, which enabled him to make observations of the fine detail of the brain's structure of unprecedented accuracy. His observations led him to question specific details of the prevailing model of Joseph von Gerlach (1820-1896), but Golgi continued to hold to the prevailing holistic "reticular model ... according to which the cerebro-spinal axis was considered to be a continuous structure, and its functions the result of a collective action."[1]

This view was soon challenged by Ramon y Cajal (1852-1934), who adopted Golgi's method of staining and produced many precise drawings of neural structures which led him to believe that, rather than a continuously connected structure, the brain consisted of discrete cells called "neurons", which communicated among themselves not through continuous connections, but by contiguity. The dispute between Golgi and Cajal continued through 1906, at which time both were awarded the Nobel Prize. Although modern research has modified details of the theory, Cajal's "neuron doctrine" is considered by many as the foundation of modern neuroscience.[1]

By 1943, enough was known about the structure and function of neurons for a mathematical description of a simple neuron to emerge. McCulloch and Pitts published a description of an artificial neuron capable of simple logical operations.[2] The McCulloch and Pitts neuron is a binary device that has a series of excitatory and inhibitory inputs The neuron functions by adding all excitatory inputs (all of which are counted equally), and if a) there are no inhibitory inputs and b) the excitatory inputs exceed a given threshold value, the neuron emits a signal, otherwise it does not.

You might wonder why a mathematical model of a simple switch that can turn on or off might be important. There are two reasons: first, mathematicians, computer scientists and logicians can build up exquisitely rich logical systems from seven basic "logic gates" which calculate simple conditions, such as "and" and "or." The idea that a brain cell can do the same thing is therefore very important to them. Second, consider that a basic switch can either turn on or off a light. Turning one light on and off might be useful as a nightlight, but consider thousands, then millions, then billions, and then billions of billions of lights. By turning some on and some off, it's possible to create a pattern. Your computer creates constantly changing patterns on its screen by turning millions of pixels on and off.

These two possibilities turn out to split cognitive scientists, computer scientists and all those who study human and machine intelligence into two subcultures[41] -- those who model cognitive processes as logical, rule governed activities, and those who model cognitive processes as pattern recognition, storage and retrieval.

The McCulloch and Pitts neuron did not learn and had to be programmed, but, once it was programmed, it could make simple decisions and, when multiples of neurons were networked together, could make interesting computations. The main value of the

---

[41] Now *there's* a coincidence -- or is it?

McCulloch and Pitts neuron, however, was not its limited utility, but that it showed that simple neurons could perform calculations.

Marvin Minsky, a graduate student at MIT, was fascinated by the McCulloch and Pitts neuron:

> Then, in Rashevsky's own journal, the *Bulletin of Mathematical Biophysics*, I found the current work of Warren McCulloch and Walter Pitts. First was the original McCulloch and Pitts 1943 paper on threshold neurons and state machines, which suggested ways to make computerlike machines by interconnecting idealized neurons. Then there was the tremendously imaginative Pitts-McCulloch 1947 paper on vision and group theory, which was the precursor of the group-invariance theorem in *Perceptrons*, the book Seymour Papert and I wrote in 1969. I'm pretty sure that it was works like these, and the flurry of ideas in the early Macy Conference volumes, that kept me thinking about how to make machines that could learn.[3]

Minsky coupled the McCulloch and Pitts Neuron with a learning rule proposed by Donald Hebb in Montreal. Hebb suggested that neurons that were frequently co-active tended to become increasingly connected:

> *When an axon of cell A is near enough to excite cell B and repeatedly or persistently takes part in*

*firing it, some growth process or metabolic change
takes place in one or both cells such that A's
efficiency, as one of the cells firing B, is increased.[4]*

In more colloquial language, this is generally
taken to mean "neurons that fire together wire
together."[42]

Along with George Miller, Minsky built a physical
machine that used about 400 vacuum tubes and forty
potentiometers that were adjusted by physical clutches
to change the probability that any given neuron would
influence another. The machine could simulate maze
learning by rats, but Minsky was not particularly
impressed with the result:

The SNARC machine was able to do certain kinds
of learning, but it also seemed to have various
kinds of limitations. It took longer to learn with
harder problems, and it sometimes made things
worse to use larger networks. For some
problems it seemed not to learn at all. This led
me to start thinking more about how to solve

---

[42] It's not clear whether Minsky took this rule from
Hebb's work, or whether he claims to have developed it
independently: "In the course of thinking about how
one might get "neural- network machines" to learn to
solve problems, I conceived of what later came to be
called Hebb synapses, after the Montreal psychologist
Donald Hebb."3.   J. Brockman, *The Third Culture:
Beyond the Scientific Revolution.* (Simon & Schuster,
1995).

problems "from the top down," and to start formulating theories about representations and about heuristics for problem solving.[3]

Nor was he much impressed by other work in the area now called "neural networks" or "connectionism": "Nothing very exciting happened in that field — that is, in the field of "general" neural networks, which included looping, time-dependent behavior — until the work of John Hopfield at Caltech, in the early 1980s."[3] He did, however, acknowledge the work of Frank Rosenblatt, like himself a graduate of the Bronx High School of Science:

> There were, however, important advances in the theory of loop-free or "feed forward" networks — notably the discoveries in the late 1950s by Frank Rosenblatt of a foolproof learning algorithm for the machines he called "perceptrons." One novel aspect of Rosenblatt's scheme was to make his machine learn only when correcting mistakes; it received no reward when it did the right thing. This idea has not been adequately appreciated in most of the subsequent work.[3]

Minsky, however, did not believe the neural network approach to modeling human intelligence was the most fruitful avenue for research:

The most important other direction in research — of attempting to set down powerful heuristic principles for deliberate, serial problem solving — was already being pursued by Allen Newell, J.C. Shaw, and Herbert Simon. By 1956 they'd developed a system that was able to prove almost all of Russell's and Whitehead's theorems about the field of logic called "proposition calculus." I myself had found a small set of rules that was able to prove many of Euclid's theorems. In the same period, my graduate-school friend John McCarthy was making progress in finding logical formulations for a variety of commonsense reasoning concepts...Soon the field of artificial intelligence began to make rapid progress, with the spectacular work of Larry Roberts on computer vision, and the work of Jim Slagle on symbolic calculus, and around 1963, ARPA — the Defense Department's Advanced Research Projects Agency — began to support several such laboratories on a reasonably generous scale.[3]

Here Minsky's deep grounding in the prevailing Aristotelian culture reveals itself in the word "deliberate." The "other direction" Minsky favors is sometimes referred to as "expert systems" or Classical Computational Theory of Mind (CCTM). Although these systems can become very complex, in principle they are composed of three systems: a set of rules, a set of facts, and an "inference engine" which applies the rules to the facts. At the root of any inference engine is a goal[43] or set of goals, such as finding the problem with an engine that won't start, deciding whether a loan applicant should be accepted or rejected, or diagnosing a medical problem.

A simple example is auto repair. Does the engine start? Yes, quit. No? Does it turn over? No? Check the battery. Yes? Check the spark. Is there spark? Yes, check the gas, if no, checks the coil...and so on.

Minsky deeply believes in rational, goal seeking actors, makes frequent reference to Freud, and the CCTM approach is distinctly goal oriented and rules based. Initially, the neural network model proposed by McCulloch and Pitts was also goal oriented -- it was designed to be able to solve particular problems. The advancements made by Frank Rosenblatt made this more evident:

The Perceptron, introduced by Frank Rosenblatt in 1958, provided a substantial advance over the McCulloch and Pitts neuron. First, the weights and thresholds for each of the input, which were identical for each neuron in the McCulloch and Pitts model, need not be identical in the perceptron, and indeed could be positive or negative, thus eliminating the need for the inhibitory synapse. Most important, the perceptron had a mechanism whereby it could learn.

---

[43] Unlike an Aristotelian system, however, the goal is never indigenous to the system, but rather imposed on it by its programmer -- in this case a mere human being and not the unmoved mover.

Each neuron in the perceptron as designed by Rosenblatt had an output of ±1. The inputs from all synapses were summed and, if the sum exceeded a given threshold, the perceptron would output +1; if not, it would output -1. The observed output was then subtracted from the desired (correct) output to give a quantitative estimate of the error, and the synaptic weights would be adjusted by a proportion of the error: if the output was too high, the weights would be adjusted to give a lower value, and vice versa if it was too low. In this way, the perceptron learned from its mistakes and did not need to be programmed like the McCulloch and Pitts neuron.

Minsky articulated his problems with the neural network approach in a controversial book *Perceptrons*[5]. In this book, Minsky and his coauthor Seymour Papert showed mathematical proof that the perceptron in the simple form proposed by Rosenblatt was incapable of modeling one of the seven logic gates -- the "exclusive-or", or XOR gate. Minsky and Papert only analyzed the simple perceptron developed by Rosenblatt, and did not formally investigate more complex multi-layer perceptrons, although they intuitively assumed these would share the same problems.

Interest in -- and funding for -- neural network research vanished abruptly after the publication of *Perceptrons*, and different observers have widely differing understandings of what has become a sore point in the history of artificial intelligence. Minsky portrays it as a minor misunderstanding:

Rosenblatt's neural-network followers were also making ambitious proposals, and this led to a certain amount of polarization. This was partly because some of the neural-network enthusiasts were actually pleased with the idea that they didn't understand how their machines accomplished what they did. When Seymour and I managed to discover some of the reasons why those machines could solve certain problems but not others, many of those neovitalists interpreted this not as a mathematical contribution but as a political attack on their work. This evolved into a strange mythology about the nature of our research — but that's another story.[3]

Robert Hecht-Nielson, on the other hand, sees it quite differently:

> The final episode of this era was a campaign led by Marvin Minsky and Seymour Papert to discredit neural network research and divert neural network research funding to the field of "artificial intelligence".... The campaign was waged by means of personal persuasion by Minsky and Papert and their allies, as well as by limited circulation of an unpublished technical manuscript (which was later de-venomized and, after further refinement and expansion, published in 1969 by Minsky and Papert as the book *Perceptrons*)[6]

The conventional account that research on neural networks diminished greatly after the publication of Perceptrons may itself be somewhat exaggerated, since some very important work was done in the 1970's, including work by the Finnish scientist Teuvo Kohonen, James Anderson at Brown, and Steven Grossberg at Boston, and the development of the back propagation model by the Harvard graduate student Paul Werbos in 1974. Why Minsky would classify Werbos' back propagation model among the "nothing very interesting" that happened before 1982 is hard to understand, since it made it possible to construct multi-layer perceptrons, which showed, contrary to Minsky and Papert's intuition, that these multilayer perceptrons *could* solve the XOR problem, which must have been a very serious blow to Minsky's credibility.

In 1982, the Caltech physicist John Hopfield published a network in which every neuron was connected to every other neuron. By varying the strengths of these connections, Hopfield's network could learn to recognize patterns. In 1987, the publication of Rumelhart, McClelland et al., *Parallel Distributed Processing: Essays in the Microstructure of Cognition,* reawakened interest in neural networks just as interest and funding in expert systems and CCTM were waning.

Among the most important developments reported in these three volumes was the Interactive Activation and Competition (IAC) neural network, in which either unconnected or randomly connected neurons are exposed to external patterns which activate some of the neurons (consider a movie marquee in which some of the light bulbs (neurons) are lit up to spell the word "cat"). The Hebb rule, of course, says that neurons that fire together wire together, so, each time the marquee displays the word "cat", the light bulbs (neurons) in the pattern "cat" become connected to each other. In this way, when part of the pattern is activated (say, "ca"), electricity will flow through the connections to the "t" and the marquee will display the word "cat." Further, if the marquee displays a picture of a cat each time it displays the word "cat", the bulbs (neurons) in the picture will be associated with the bulbs (neurons) in the word, so that when the word is displayed it will call up the picture, and when the picture is displayed it will call up the word.[44]

Although all neural networks share certain characteristic properties, the IAC network, along with Kohonen's self-organizing maps, differ from perceptrons in a fundamental way. Perceptrons, including multi-layer back propagation networks, are

---

[44] The connections among neurons in an artificial neural network are, of course, just numbers, and these numbers can be interpreted as distances; highly connected neurons are "close" to each other, and loosely or negatively connected neurons are "far apart." Teuvo Kohonen capitalized on this analogy so that his technique, which is similar to the IAC network, produces "self-organizing maps." 7.   T. Kohonen, *Self Organizing Maps*, 3 ed. (Springer, Berlin, Hedelberg, New York, 2000).

purposeful machines. They are designed to model complex patterns of events so that they can make predictions about future events. To do this, they "study" cases in which the outcome is already known, and "practice" until they can predict these known outcomes to within a tolerance specified by their programmer. One they have "learned" the training set, they can make predictions about future events.

A baseball game is a good example. At any given game, we can note the temperature, rainfall, place of the home team, place of the visiting team, the number of tickets sold, the number of hot dogs consumed, the amount of beer sold and the number of fans arrested. The first four of these are considered inputs, and the last four outputs (they need not be equal, and it is better to have more inputs than outputs).

Then we take a game that has already been played. We take note of temperature, the rainfall, the place of the home team, the place of the visitors, and assign these values to the input neurons. The input neurons are randomly connected to a set of middle or "hidden" neurons[45], which, in turn, are randomly connected to the output neurons, one for each output variable. The values of the input neurons will travel through the random pathways to the hidden neurons, which will in turn send the signals to the output

---

[45] These neurons are called "hidden" because they have no connections outside the network itself, and neither receive nor send information to or from the environment.

neurons, which will be activated at some (random) level. These outputs will be compared to the correct outputs -- the actual number of tickets sold, hot dogs consumed, beer drunk and fans arrested -- and will, of course, be wrong.

But they will *be wrong by some amount*, and this error can be expressed as a function of the connection strengths. By a well-known technique from the calculus, the connection strengths will then be changed, and the next case will be read and the process repeated. After (typically thousands) of passes through the data, the network will usually learn the data. Then one can enter the predicted temperature, rainfall, standing of the home team and the visitors for any given future game, and the network will predict the number of tickets sold, hot dogs consumed, beer drunk and fans arrested. In general, this usually works better than standard statistical techniques like linear regression, but generally not enough better to convince many social scientists to switch to the newer technology.

Shortly after the PDP volume's publication, William Richards spent a semester on sabbatical leave at the University at Buffalo. Richards wrote the first social network analysis software, NEGOPY, while he was a graduate student at Michigan State University in the late 1960's and early 1970's, and continuously improved and developed it until his untimely demise. He was an excellent FORTRAN programmer, and together he and I wrote a back propagation program as a second generation SPOT. The input neurons for SPOT consisted of a 40x8 matrix of ones and zeros. Each column of the matrix represented the ASCII code for a symbol. This enabled SPOT to accept ordinary language phrases up to 40 characters long as input. SPOT's output nodes were a mirror or the input nodes, so SPOT's output consisted of forty character ASCII symbols.

Spot is trained on pairs of phrases, the first phrase being the input and the second phrase being the desired output:

Input: Hello, Spot.
Output: Hello, Joe.
Input: How are you, Spot?
Output: I'm well, thank you.
Input: How's your mother, Spot?
Output: I'm a machine, Boss.

When SPOT reads this input file, a small subroutine converts the phrases into their 8 bit ASCII codes, which become SPOT's input neuron values. SPOT then feeds forward these values through the initially random links to the hidden and then output nodes (which are then converted back to ACII characters.) At first, these outputs are random ASCII characters, but, through the back propagation error correction, eventually they converge on the right answers.

The following training session shows that the first input phrase, "Hello, Spot." produces a random burst of ASCII characters. The second input phrase,

"How are you, Spot." yields the output "Hello, Joe.", which would have been the correct output for the previous phrase. SPOT just learned this response in the previous cycle, and it is the only thing it knows, so it says it. As SPOT cycles through more learning cycles, the output improves (notice it begins to get some of the letters right) until it finally converges on the correct answers:

Hello, Spot.
_!/ 5!A!/' /!' '

How are you, Spot?
Hello, Joe.

How's your mother?
Imnhw+$åxc6`,h po1

Hello, Spot.
I'm w$lnk`vja,`@oos*

How are you, Spot?
Igm w$ljc`~ha,` oos*

How's your mother?
I'm w$ljc`~ha, oos*

Hello, Spot.
I'm o$ljc`nja, oos"

How are you, Spot?
I'm o$ljc`nha, oos"

How's your mother?
I'm o$ljc`nha, oos"

Hello, Spot.
J+mlo,,äs`.h, opp"

How are you, Spot?
I'm o$ljc`nha, oos"

How's your mother?
I'm o$ljc`nha, oos"

Hello, Spot.
Hmklo, Joe. (_`0

How are you, Spot?
I'm o$ljc`vha,` oos"

How's your mother?
I'm o$lnc`nha, oos"

Hello, Spot.
Hello, Joe.

How are you, Spot?
I'm o$ljk`nha,` oou"

How's your mother?
I'm a machine, Boss!

Hello, Spot.
Hello, Joe.
How are you, Spot?
I'm well, thank you.

How's your mother?
I'm a machine, Boss!

Now that SPOT is trained, it's possible to converse with it. If given the exact phrase it learned, it will always respond with the correct answer, as the first three exchanges show:

> Hello, Spot.
> Hello, Joe.
>
> Hello, Spot.
> Hello, Joe.
>
> How are you, Spot?
> I'm well, thank you.

But if the input is too far off from what SPOT learned, it will not respond correctly:

> How are you?

Notice there is no answer, because a small subroutine prevents SPOT from saying anything if the output has a large random component. But if the input is not too far off from what SPOT learned, it can get the right answer anyway:

How are you, Spit?
I'm well, thank you.

How's your mother, Spot?
I'm a machine, Boss!

How's your mother
I'm a machine, Boss!

Finally, "How you doin'?" is just too far off, and SPOT remains silent:

How you doin'?

Unlike typical back propagation models, SPOT was not designed with any practical purpose in mind, but rather as an aid to the study of the properties of neural networks. One of the first things to strike an observer running SPOT for the first time is its superficial resemblance to human behavior. First, it can not only recognize a learned phrase and respond with the correct answer, but it can recognize a somewhat distorted version of the phrase it learned, even though it has never encountered it before, and respond with the correct answer.

After spending more time with SPOT, one sees that, like a human being, SPOT can be either under or over trained. If SPOT reviews the training set too few times, its answers are imprecise, with spelling errors and wrong choices. If over trained, SPOT performs flawlessly when given the exact inputs it learned, but loses the ability to respond correctly when the input is different -- it loses its ability to generalize to new inputs.

Spot makes mistakes like human beings make when they are first learning. Rumelhart and McClelland showed that, when learning the past tense of English words, simple artificial neural networks frequently apply the usual rule for past tense by adding "ed" to the present tense of even irregular words, such as "hurt", "hurted." As learning progresses, the network gets the irregular verbs right. They note that this is similar to the way children learn.[8] This is a very controversial point, because it directly challenges the standard (Aristotelian) model of language, which assumes that language is fundamentally rule governed behavior. Of course, in a connectionist system, there are no rules, only patterns that match still other patterns or not.

Defenders of establishment linguistic theory responded in depth. In a lengthy paper, Pinker and Prince[9] provide what at first appears to be a complete rejection of Rumelhart and McClelland's argument:

> We analyze both the linguistic and the developmental assumptions of the model in detail and discover that (1) it cannot represent certain words, (2) it cannot learn many rules, (3) it can learn rules found in no human language, (4) it cannot explain morphological and phonological regularities, (5) it cannot explain the differences between irregular and regular

forms, (6) it fails at its assigned task of mastering the past tense of English, (7) it gives an incorrect explanation for two developmental phenomena: stages of overregularization of irregular forms such as bringed, and the appearance of doubly-marked forms such as ated, and (8) it gives accounts of two others (infrequent overregularization of verbs ending in t/d, and the order of acquisition of different irregular subclasses) that are indistinguishable from those of rule-based theories. In addition, we show how many failures of the model can be attributed to its connectionist architecture. We conclude that connectionists' claims about the dispensability of rules in explanations in the psychology of language must be rejected, and that, on the contrary, the linguistic and developmental facts provide good evidence for such rules.[9]

But, after a painstaking assessment of the linguistic structures and processes that Rumelhart and McClelland's very simple one-layer perceptron can't model, they conclude by acknowledging that a more advanced artificial neural network would be fully capable of modeling any rules based system:

> We do not doubt that it would be possible to implement a rule system in networks with multiple layers: after all, it has been known for over 45 years that nonlinear neuron-like elements can function as logic gates and that

hence that networks consisting of interconnected layers of such elements can compute propositions (McCulloch and Pitts, 1943). Furthermore, given what we know about neural information processing and plasticity it seems likely that the elementary operations of symbolic processing will have to be implemented in a system consisting of massively interconnected parallel stochastic units in which the effects of learning are manifest in changes in connections. These uncontroversial facts have always been at the very foundations of the realist interpretation of symbolic models of cognition; they do not signal a departure of any sort from standard symbolic accounts.[9]

Here Pinker and Prince (correctly) say that connectionist systems can be devised that can model any rule based model, and rule based models can be conceived that can model any connectionist system. After all, none of the connectionist systems discussed in this debate are connectionist systems after all, but all are in fact simulations of connectionist systems implemented in rule based serial computers. The idea that Rumelhart and McClelland's primitive early model is insufficient to model human linguistic behavior perfectly is obviously true, but trivial. And, since Pinker and Prince acknowledge that a more sophisticated connectionist system can model human linguistic behavior, why are we having this argument? *The argument is really the clash of the two cultures, Ionian and Athenian, in modern dress.* Whether people describe processes as rule governed or law governed is a good indicator of which culture they belong to. After all, the Hebb rule -- neurons that fire together wire together -- is a rule when implemented in a computer program, but is a causal process when implemented in a brain.

Connectionists responded to the Pinker and Prince argument mainly by building more sophisticated networks that produced increasingly accurate models[10], and, because of their usefulness in modeling brain impairment, less accurate models.[11] In fact, a remarkable array of specialized neural architectures emerged over the next decades. At Buffalo, SPOT -- already a much more sophisticated model than Rumelhart and McClelland's early network -- evolved into ROVER with a very simple change that made its behavior radically more complex.

While SPOT exhibited several human-like characteristics that still amaze the first time viewer, certain aspects were very machine-like. For example, SPOT has no self-awareness whatsoever -- it doesn't know what it has said. That makes it possible for obviously mindless repetition:

> Roll over, Spot.
> OK.

> Roll over, Spot.
> OK.

> Roll over, Spot.
> OK.

But with a very simple modification, this changes in a radical way. Adding a feedback loop between SPOT's output neurons and half of its input neurons means that part of the input on any given cycle will be SPOT's own behavior on the previous cycle. In this sense, ROVER (the modified SPOT) is "aware" of its previous behavior. This has several profound consequences. First, it's possible to train ROVER to have this conversation (which is impossible for SPOT):

> Roll over, Rover.
> OK.
> Roll over, Rover.
> Again, Boss?

Adding a simple feedback loop into SPOT changes it into what is called a *recurrent neural network* (RNN). Recurrent neural networks are very powerful indeed, one of the most powerful of all being the human brain. Although ROVER is about as simple a recurrent neural network as can be -- a single feedback loop allowing ROVER to be aware of only the last thing it said -- it makes ROVER's behavioral range so much larger that it is no longer feasible to train it manually, but a computer program had to be written to write ROVER training files, because it is so difficult to predict what ROVER will say for any given input. Because ROVER's last output is always part of its next input, it is possible simply by adding no new inputs to have ROVER recite a sequence of arbitrary length. ROVER can recite "A Bicycle Built for Two," for example.

Because these networks are tested against a set of known answers, they are usually called *supervised* neural networks. As interesting as they may be, even more fascinating are networks that learn by themselves. IAC networks and self-organizing maps, on the other hand, use a completely different learning rule -- the Hebb rule. They are not tested against correct answers, and are not seeking any goal or avoiding any penalty. They simply recognize and store any patterns of information to which they are exposed.

Historically, the first idea for artificial neurons, of course, was the McCulloch and Pitts neuron. The combination of several neurons into a vector or matrix (or *layer*) was a natural extension of the notion of a single neuron, and the idea of adding extra layers still an obvious further extension. But Hopfield's idea of having just a bunch[46] of neurons which are either unconnected to each other or randomly connected to each other and allowing the connections to emerge over time is a much better idea. Building, say, a traditional three layer back propagation network, for example, means that neurons in any given layer are prohibited from communicating with any other neurons in the same layer, and that neurons in any layer can only be connected to the next layer forward. The likelihood that such a configuration is the optimal configuration to model any actual process is vanishingly small. Hopfield's model, however, allows for the possibility that the network will develop the set of interconnections among neurons that is optimal for whatever task it is assigned.

Hopfield's original network allowed only binary activation values for the neurons -- either 0, 1 or -1, 1 -- which severely limits the information carrying capacity of the network.[47] Allowing the neurons to take on continuous values,[48] however, and adopting a Hebbian

---

[46] Not a technical term.

[47] The number of patterns a binary Hopfield network can store and retrieve is a little more than fifteen percent of the number of neurons it has. Assuming there are around 64 billion neurons in a single human brain, that yields about 10 billion patterns, which is not that shabby.

[48] In organic neural networks, neurons fire at a constant amplitude, but with variable frequency, with more activation yielding higher frequency pulses. In an artificial neural network, increased intensity is, of course, represented by a higher number.

learning rule, produces a class of networks generally called *unsupervised* or *self organizing* networks.[7, 8, 12] Such networks reshape themselves to resemble their environment.

Consider the very simple unsupervised neural network in *Catpac*, a popular program for text analysis. Catpac is similar to the IAC networks described by Rumelhart and McClelland, but modified to work well as a text analysis program. Catpac reads text by passing a moving window through the text. The default window is seven words long, so Catpac reads the first seven words in a text, then slides the window one word to the right and reads words two through eight, then slides again and reads three through nine, and so on until it has scanned the entire text once.

Catpac assigns a neuron to each major word in the text (minor words, such as articles, prepositions and the like, can be excluded if so desired) and activates the neuron when its word is in the sliding window. At the beginning of the first cycle, the first seven words will be in the window, and all seven will be activated. Because they fire together, they will wire together, and so the connections among them (originally zero) will all be incremented by a small amount. The window then slides one word to the right, and words two through eight will be in the window. These seven words will be activated, and their interconnections incremented. This continues until all words have been read.

But at the end of each cycle (a cycle is one window), every neuron in the system -- not just the ones in the window -- will be polled. As the window moves through a larger proportion of the words, many of the neurons will be interconnected, so when words in the window are activated, they may spread their activation to other neurons to which they are connected, even if they are not in the window. How active they will become depends on how many active neurons they are connected to, and how tightly they are connected. Finally, all the neurons activations are reduced by about 90 percent[49], and all connections are reduced by a very small amount to represent forgetting, and the remaining connections are normalized and the average of all connections is subtracted from all connections, so that the least connected will have negative (inhibitory) connections while the most connected will be positive.

The result of these very simple processes is a word by word matrix of numbers which represent the extent to which words "go together" in the text. Patterns of words that occur frequently in the text will cluster together, while words that seldom are present in the same pattern will be far apart. Thus, whatever patterns of words occur in the text will be represented in the pattern of interconnections among the neurons.

---

[49] When a word is in the window, Catpac is "looking" at it. When it falls out of the window, Catpac may "remember" it. But if it remembers it perfectly, it will be forever "seeing" everything it has ever seen, which would, of course, be a bad thing. That's why the neuron representing the word must have its activation drastically reduced as soon as Catpac is no longer looking at it.

Unlike the perceptron, feed forward and back propagation models, this unsupervised model begins with no structure at all, with the number of neurons and the patterns of interconnections among the neurons completely determined by the inputs it experiences. Any neuron can act as an input, hidden or output node on any cycle, and the path length from one node to another is completely determined by its experience, not by the programmer. Since the network cycles and information can travel backwards as well as forward, the network is a recurrent neural network (RNN) like Rover. Since the neurons can take on continuous activation values, the information carrying capacity is much larger than the dichotomous Hopfield network.

Learning is only one of two ways that neural networks can develop an optimal structure. Over the long term, the most powerful method of modifying the structure of neural networks is natural selection. In organic neutral networks, natural selection has led to the construction of highly specialized neural structures that had survival value. These specialized structures are the mechanism underlying Mach's understanding of how the senses' built in *a priori* concepts were at the time of their evolution *a posteriori*.

1.   G. Cimino, Physics Riv Int Stor Sci **2**, 41 (1999).
2.   W. McCulloch and W. Pitts, Bulletin of Mathematical Biophysics **7**, 18 (1943).

3. J. Brockman, *The Third Culture: Beyond the Scientific Revolution.* (Simon & Schuster, 1995).
4. D. Hebb, *The organization of behavior. A neuropsychological theory.* (Wiley, New York, 1949).
5. M. L. Minsky and S. A. Papert, *Perceptrons.* (MIT Press, Cambridge, MA, 1969).
6. I. S. N. Berkeley, (1977).
7. T. Kohonen, *Self Organizing Maps*, 3 ed. (Springer, Berlin, Hedelberg, New York, 2000).
8. D. E. Rumelhart, McClelland, James L., *Parallel Distributed Processing: Explorations in the Microstructure of Cognition.* (MIT Press, Cambridge, 1982).
9. S. Pinker and A. Prince, Cognition **28**, 120 (1988).
10. J. A. Bullinaria, 1994.
11. M. S. C. Thomas and A. Karmiloff-Smith, Psychological Review **110** (No. 4), 35 (2003).
12. S. Grossberg and G. Carpenter, Applied Optics (1987).

Chapter 22: Meanwhile, back in Wisconsin...

And now, research with artificial neural networks and Galileo spatial models both indicate that human mental processes do not appear to be hierarchical, syllogistic processes as Aristotle's model holds, but rather network processes, where concepts mutually influence each other through a network of synaptic interconnections.

In the beginning, we discussed the Wisconsin Dow Demonstration as an example of a concept whose origin is well known. The concept of the Dow Demonstration at first existed only in the brains of the participants, but later spread through media and word of mouth until it existed world wide in many brains and artifacts. One of those artifacts that still exists in the University of Wisconsin Archives is a transcript of an Oral History of the Dow Demonstration by William H. Sewell, then the Chancellor of the University. Although Sewell has now departed this mortal coil, Catpac can read his oral history and uncover the main patterns that represent his concept of the Dow Demonstration.

Since Catpac's artificial neural network represents these patterns as numbers that can be considered distances among the neurons, it's possible to enter Catpac's neural network into the Galileo computer program and portray the concept visually in space.

Figure 10 shows Sewell's concept of the Dow Demonstration as it is represented in the oral history. The first thing to strike the eye is the remarkable distance between Sewell's self-conception ("me" on the far right in the picture) and the rest of the concepts in the Dow Demonstration. This distance is so large that it crowds all the other concepts together and makes them difficult to read. Clearly, Sewell sees himself as widely separated from the faculty (top left) and the students (bottom left), and the faculty and students also widely separated from each other.

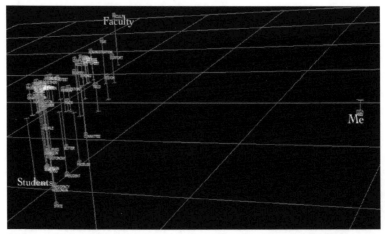

Figure 10: Galileo rendering of Catpac analysis of William H. Sewell oral history

The crowded cluster on the left represents mainly people and structures in the university administration and faculty, from which Sewell is also quite distant. To a large extent, Sewell, who is after all a sociologist, sees the affair in terms of the roles of the people involved and the social structures which guided the events. He contrasts the governance structure of the University of Wisconsin with that of the University of Michigan. Wisconsin, like Berkeley and unlike Michigan, has a leadership divided between a president and a chancellor, which weakens the hand of the chancellor. He also sees that, as a new chancellor with a staff left over from the previous administration, he neither knows well nor can count on the loyalty of his organization, and does not know who called the Madison police. He sees himself at the crossroads of forces from liberal and conservative faculty and a disorganized collection of radical students competing with the faculty, the administration and each other for power. He sees his power as limited, lacking the authority to call off the Madison police. Although Sewell denies that his sociological expertise was of any help to him during the demonstration, his analysis of what happened is a highly skilled sociological analysis.

Of course, this is a superficial analysis of Sewell's concept of the Dow demonstration consisting solely of a computer analysis of part of a single document, but the point of the example is not to provide an exhaustive understanding of Sewell's beliefs and attitudes. This might be the object of a larger dissertation, or a book in itself. But the larger point is that individual's concepts can be modeled by artificial neural networks, because those concepts are real biological neural networks, and further, that any neural network can be modeled as a Galileo space. The connections among neurons can be modeled as distances, and those distances can be projected into a mathematical space. The aggregate cultural concepts can be modeled as the aggregate of the individual models. Once again, in the Ionian tradition of science, these are *models*, not descriptions or explanations, and are neither true nor false, but they are precise, useful models that fit our experience to much more precise tolerances than the Athenian categorical model.

The simple unsupervised neural network in Catpac can show clearly how the comparative Ionian measurement model differs sharply from the Athenian categorical model. In Aristotle's syllogistic, objects belong to categories because of their essential attributes. Thus, every rational animal -- that is, a being that has all the essential attributes of animals plus the attribute of rational thought -- is a man. (Women, of course, were not thought to be rational by Socrates, Plato and Aristotle.) The notion of *degree* of rationality or *degree* of animalism has no meaning in this system of thought. Every man has exactly the same amount of "man-ness", just as a German Shepherd and a Chihuahua have the same amount of "dog-ness."

In a neural network, however, objects are grouped together based on the extent of their interconnectedness, and the notion of a sharply bounded category is brought about only by the notion of *thresholding*. Indstar[1] is a variation of Catpac which uses the identical neural model, but has a different form of input and output. Rather than reading a text, parsing it into individual words, and sliding a window through those words, Indstar reads lists of lists. Each list can be thought of as a Catpac window, although, unlike the

Catpac window, the lists need not be the same length. Consider a number of customers at a grocery store. The items each customer buys is a list, and the list of all the customer's lists would serve as an input to Indstar.

Indstar reads the first list, creates a neuron for each item on the list and activates it. It then reads the second list and continues through the lists just as Catpac reads through all the windows. At the end of the last list, Indstar's neural network remembers how much the items on the lists "go with" each other, because those items that tended to co-occur on many lists will be more tightly interconnected than those that co-occur seldom or never. (In Galileo space, those that co-occur frequently and are thus tightly linked will be closer to each other than those that co-occur seldom or never co-occur, and those which seldom or never co-occur will be further apart.)

After Indstar has learned the interconnections among the neurons representing the items on the lists, it is possible to query the system. The user simply activates one or more neurons by entering their name, and Indstar activates the other neurons that are sufficiently tightly connected to those activated by the user. In Figure 11, the user has named two cars, and Indstar then activates the other items -- in this case, cars -- which are most closely connected to those activated by the user:

```
Enter as many items as you want, Your Uselessness.(Ctrl z when done)
AUDI
Enter as many items as you want, Your Uselessness.(Ctrl z when done)
PORSCHE
Enter as many items as you want, Your Uselessness.(Ctrl z when done)
Z
Do you want these values clamped?
                        CAMARO              Activation level = 0.0050
                        MUSTANG             Activation level = 0.0050
                        AUDI                Activation level = 1.0000
                        PORSCHE             Activation level = 1.0000
                        CORVETTE            Activation level = 0.0025
```

Figure 11: Cars connected to Audi and Porsche according to Indstar

Notice that, unlike the Aristotelian syllogistic model, membership in this "category" is not absolute; the neurons' activation levels show that Corvette is only half as much a member as are Mustang and Camaro.

This is not the only way in which the neural network model diverges from the syllogistic. Neural networks are not hierarchical. A hierarchical network consists of a category and its members. Each of the members of the category may themselves be categories with members, and so on *ad infinitum*. But the pattern of interconnections determines which set of items will "belong" to a given category or cluster given that some subset of neurons are activated, which means that category membership is not only relative and variable, but context sensitive. Thus, when "PONY", "THOROUGHBRED" and "COWBOY" are activated, Indstar produces a completely different "category" that includes the term "MUSTANG." Notice that no cars occur in this equine realm.

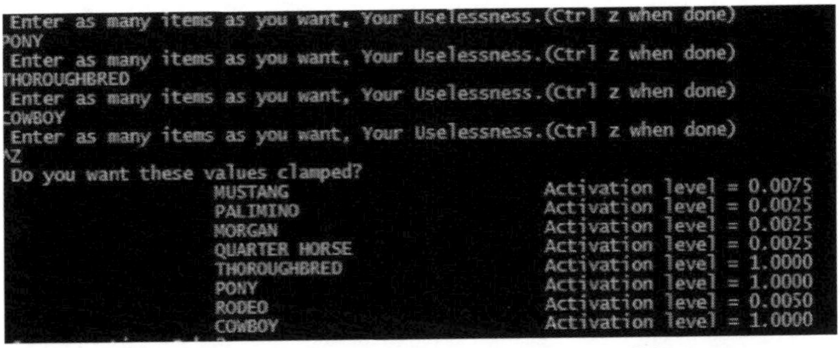

Figure 12: Indstar sees MUSTANG as a horse in this context

None of these simple programs, built of only hundreds or, in some cases, thousands of neurons can be compared even remotely with the 64 billion neurons in the human brain, arranged through hundreds of millions of years of evolution into specialized complex structures selected by ever changing environmental conditions to become the remarkable thing that is a human brain. And, of course, these are not really neural networks, but rule governed serial computers emulating neural networks. But the artificial neural networks and the organic human brain share the same DNA -- both are neural networks, and share the same principles of operation:

First and foremost, neural networks are approximate and uncertain. While they can come to recognize patterns in their environment, they are always subject to error. Neural networks are generally not good at arithmetic, and the square root of 69 is "eight something." Perfect truth forever eludes the neural network, but, over time, it can build up increasingly precise and useful models which are neither true nor false, but useful. Models can be elaborated, refined, modified and occasionally, replaced, gradually or abruptly.

The knowledge of neural networks is comparative and not categorical. Items cluster together on the basis of their perceived similarity, and not because they all share equally some essential feature. Reasoning is comparative rather than deductive. Leopards, panthers, jaguars and housecats are not grouped together because they all share the essence of cat-ness, but rather because they are all built on the same neurons and share substantial amounts of patterning, and thus are more similar to each other than they are to anything else.

Neural networks can pursue goals, but ordinarily do not. Neural networks can be structured to seek goals -- recently, a neural network trained to recognize highway signs won first place in a sign recognition contest in which the second place finisher was a team of human beings. But unsupervised networks like Catpac or Indstar are not seeking goals, but rather matching patterns. When SPOT says "Hello, Joe" in response to "Hello, Spot," it is not trying to ingratiate itself, nor does it want to please the other -- it is simply finding the best matching pattern.

The internal organization of any unsupervised neural network is always based on the organization of its environment. The patterns encountered in real time are internalized in the neural network as connections among its neurons, and these patterns are a model of the environment of the network.

Finally, a network of networks is a network, or, more colloquially, two heads are better than one.

Each of the simple neural networks we've seen in this book is capable of some human-like behavior. But to expect them to do everything an actual human brain can do is too much. These simple models have hundreds of neurons and thousands of connections. The human brain, after eons of natural selection, has about 64 billion neurons and perhaps between one and five thousand times that many connections (no one really knows for sure yet). Moreover, specific structures have evolved that make humans what they are today.

We also need to recall that none of the neural networks we've studied here are actually neural networks -- all are simulations of neural networks in serial digital computers. While computers calculate much faster than humans, and while electronic messages travel at the speed of light while nervous impulses travel only about 100 meters per second, thousands of billions of individual streams of nervous impulse can travel simultaneously in the massively parallel brain. Not only does this make the brain faster than a serial computer, but it makes it even more capacious, since the order in which different message streams are received can be taken into account to represent different patterns.

All the effects we've seen in Galileo research are processes that can be accomplished by neural networks. Because of the interconnected character of the brain, effects can trigger other effects which can trigger still other events. Attitudes can change, which triggers other, related neurons which can "counter argue" and create oscillations like those observed in Galileo research.

The rule for scientists, of course, is that every theory is tentative, subject to revision or rejection at any time. But for the other culture, the Athenian culture, different rules apply. In Aristotle's rhetoric, the burden of proof is on the affirmative, but I think it's safe to say that we've established a *prima facie* case that all the behaviors of which human beings are capable are products of the secular action of human brains and networks of brains, and can be accounted for by the methods of Ionian science. The burden of rejoinder rests with the negative.

1.   B. Battleson, H. Chen, C. Evans and J. Woelfel in *Sunbelt XXVII, INSNA Social Networking Conference* (St. Pete Beach, FL, 2008)

Chapter 23: A Good Start

And now, after dozens of years of research by dozens of scientists, the possibility of genuine Ionian science of human phenomena seems real. Perhaps we can begin to push back the veil of ignorance and superstition that obscures our understanding of who we are...

We began this book with an inquiry into the differences between the cultures of two distinct social networks -- the Ionians and the Athenians. The Ionians' notion of knowledge and understanding is a symbolic model that represents our experience to ever-smaller tolerances, but a model that is always tentative, always subject to modification in the face of new observations, and always subject to replacement when a better model is available. Their method of observation is always comparative, never absolute. They recognize only one world -- the world we observe -- which can be understood without recourse to supernatural ideas. There is no underlying purpose for the world. The main intellectual fruit of the Ionian model is science.

In sharp contrast, the Athenians believe that knowledge is perfect, absolute and unchanging. The world we observe is not the source of knowledge, which comes from supernatural sources -- the world of ideas, or the uncaused cause. Perfect knowledge comes from knowing the causes of things, which flow back in an unbroken chain to the creator of all things, the uncaused cause or unmoved mover. Observation is categorical and reasoning is syllogistic. All action, including human behavior, is teleological -- everything is done for a purpose. Perfection is for the heavenly bodies; the earth is an imperfect place and can only be known imperfectly. The main intellectual fruits of the Athenian model are philosophy and theology -- and literature.

Our historical survey from the Fifth Century B.C.E. and beyond to the present day shows that the members of these two social networks seldom interact, and that the social network of modern social scientists has very little contact with the network of Ionian science indeed. Contemporary social science -- at least the so-called "quantitative" branch -- has tried to construct a science using Athenian tools: categorical scales, correlation and the search for "truth" through a statistical model of falsification.

We've seen that both the Ionian model and the Athenian model are capable of generating many, many theories. But there the similarity ends. Inexorably, inadequate theories are decisively rejected in the Ionian model, while in the Athenian social sciences, contradictory theories live side by side forever. We've examined and rejected the simple explanation for the relative lack of success of the social sciences: that human phenomena are too volatile and evanescent to be studied scientifically, or that they are just harder to study.

We've also examined and rejected the idea that human phenomena require a special kind of science, one based not on the comparative measurement model with replication of Ionian science, but on a crude categorical scaling model along with non-metric measurements, correlations and statistical significance tests. We've gone the extra mile and actually tried to construct a scientific model of human cultural and cognitive structures and processes using *only* comparative measurements and we find it works well.

Briefly, the Galileo model defines concepts as bundles of interconnected neurons. These bundles are themselves interconnected with still other bundles to make higher-order concepts (bigger bundles). The neural network of interest does not live in individual brains, but is the global network of all human brains. These concepts, at any level, can be modeled as points in a high dimensional non-Euclidean space. The distances between any pair of concepts are proportional to the degree of interconnectedness between them, although the exact functional form of this relationship is not known. Changes in the level of interconnectedness correspond to movements in the Galileo space.

Each concept is defined in Galileo space by its position vector (technically, a contravariant tensor). Combinations of concepts are themselves concepts whose meaning is given by the average of their position vectors. When two concepts are associated in a message (i.e., when their neurons are simultaneously activated) their interconnectedness is increased, and this is modeled in the space by their moving closer together.

A very special concept in the space is the self. The self is the organizing focus for the individual person, although a clearly defined self for the collective consciousness has not yet emerged. Objects closest to the self are most closely associated (highly connected) with the self. This makes it possible to engineer "messages" that move objects of interest closer to or further from the self, which is one basis for the engineering capability of the Galileo model.

The conceptual structure or culture of the collective is generated in the long term by natural selection and in the short term by Hebbian learning. In this way, the collective consciousness develops a simulacrum of its environment in the interconnections among its neurons. This is not a linear representation of nature as is, but rather a systematically adjusted picture that, over generations, has survival value.

What does "works well" mean? It means that it models important processes of interest, such as attitude formation and change, the development of concepts, the detection, storage and retrieval of patterns in our environment. It means that the model is subject to constant revision and change in the face of observations. It means that it has developed a powerful and useful engineering capability -- all of which have eluded standard social science models. It is a model that is consistent with the neural infrastructure of society.

The model Galileo works well, provides a solid basis for measuring and engineering cultural processes, and may even be, as the Rand Corporation says, "... the closest that any social science approach came to providing the basis for an end-to-end engineering solution for planning, conducting, and assessing the impact of communications on attitudes and behaviors."

But it is, after all, only a theoretical model, and, in Ionian science, these come and go. The real purpose of this book is to show that human cognitive and cultural processes can be studied scientifically. It's not even necessary to use a watered-down semi-scientific method. Human beings and their societies are a part of nature. They are not special. They are not inscrutable, unmeasureable or in any way privileged. Of course they are hard to study, but science is difficult. But, if you could construct a social science using exactly the same methods as physical science, why would you not? And, if you built a special social science using different methods that didn't work, how long would you persist?

It is true that we haven't made the kind of progress in understanding ourselves that we have in understanding our universe, but the reason for that is our cultural prejudice inherited from our Greek past. The outcome of any study depends not only on what is being studied, but also the way in which it is studied. We have studied ourselves with completely different methods than the ones we use to study the world we live in, and, unsurprisingly, have come to a different result.

When we study *any* topic with categories and syllogisms, the result is a set of nested categories. Within these categories, objects are undifferentiated from each other. When we study any topic with ratio comparisons to a standard, we end up with space. The character of that space will depend on what we study, and, in the end, will turn out to be a model of the phenomenon of interest, capable of infinite refinement.

Measurement means comparison to some standard. There are no entities in the universe that inherently require a different kind of measurement. If there is one thing this book hopes to establish, it is that all things human can and should be measured this way. When human processes are studied as continuous variables, the lacunae within categories vanish, and events can be described precisely. It should go without saying that things that have not been described can't be explained, but apparently it doesn't. There are mountains of conflicting explanations for human phenomena that have not yet been seen clearly, since we have looked at them through only the crudest categorical schemes.

Galileo theory and method have given us a clearer glimpse of human cognitive and cultural processes, but it is as yet only a glimpse. After all, it is only the work of a few dozen people over a few dozen years, and we had to learn scientific method as we went. We are still not very good at it. But -- hopefully -- we have set an example of how to see more clearly. First, we must believe that it is possible. Starting out by believing that human activities are inscrutable assures failure. And we must believe it is *worth doing*. The level of commitment and resources devoted to the understanding of ourselves is quite puny in contrast to the level of commitment we've made to the study of the world we live in.

We need to give up the idea that we already know the answer. Ancient and archaic notions about human beings -- that everything we do we do for an end, that we are fundamentally free to choose, that we are imperfect beings that can't be clearly understood -- these are not facts about people, but legacy code left over from Aristotle.

We must stop trying to "falsify" our theories and start trying to modify them to fit observations -- we need to stop asking if our hypotheses are false or not, and start measuring how far off they are and how they need to be changed to be right. Error needs to be our guide, not our enemy. This requires comparative measurement. That's what science does. Doing something else and calling it science is not working, and can't work. We need to avoid the categorical judgment of whether a theory is false or not -- they're all false. We need to compare them to each other to see which are closer to matching observations, and this requires comparative measurement.

We need to hold social science to the same standards as any other science. Yes, Aristotle said that educated persons ought to expect only the level of precision appropriate to the subject, but Aristotle was wrong. If it doesn't fit observations, it's wrong. And the answer is not to have so much imprecision that anything can fit to within measurement error. Precision must constantly increase.

And we need to change the way we check our work. Science is the process whereby we make observations and communicate them to others, who must check them. The idea that we can check them ourselves by looking up probabilities in a table is contrary to the nature of science, and has clearly failed. We, literally, need to make observations and communicate them to others, who must check them. Studies must be replicated; important studies need to be replicated again and again.

Science can understand humanity in the same way as it "understands" anything else: by constructing ever more precise and encompassing models. Galileo is a very early model. But it is a model made entirely of comparative concepts and measured entirely by comparative methods. It works, "pretty nearly." It's a good start.

The development of a truly scientific social science is a large goal, but only a small part of a much more important quest. We need to understand ourselves. As it is, the culture of Western society is fundamentally Athenian. Not only social scientists, but we as a collective culture and as individuals understand ourselves in a categorical way. We are as a whole and as individuals immersed in the Athenian myths -- that everything we do, we do for a purpose, that we are volatile and evanescent, fundamentally "fuzzy" and unobservable, subject to aleatory factors, that we move in discrete jumps from one state of being to another, with the intervening process blurred. Our economic and political models are based on these false premises, and, not knowing who we are and how we work, we blunder darkly from one crisis to another.

A hundred years ago, Charles Spearman announced that "G [intelligence] is in the normal course of events determined innately; a person can no more be trained to have it in higher degree than he can be trained to be taller."[1] But Spearman was wrong. Any kind of computing machine, whether silicon or carbon based, depends not only on the speed and configuration of its hardware, but on the way it is programmed. The Athenian Culture is software that resists change by blinding itself to its environment. The Ionian Culture maximizes our ability to communicate with the world we live in. It enhances our ability to observe. It makes us smarter.

As physical science has progressed, it has learned to measure "mysterious" things like temperature, time, motion and change -- and we've seen that these things are indeed mysterious in a categorical world. As science collectively learns to observe these things, they diffuse out into the culture and, eventually, into individual brains. That's why we have schools, and why we teach kindergarten children how to measure distance and time, and why we print tables of conversion on the covers of school notebooks. Developing a scientific way of describing ourselves and our actions will bring about massive change in our collective and individual self concepts. The Buddha was right; we need to be aware. We need to be self-conscious.

One more thing. Richard Feynman, who has guided us on this journey from the beginning, added one more criterion to successful science: lack of respect.

No one must be considered beyond checking; everyone makes mistakes. And so we pay Dr. Feynman one last measure of respect when we say that he, too, makes mistakes. In a speech he made at Caltech's commencement in 1974, he discussed "sciences that aren't science." Before he began to list some of the most important problems, he said, in an uncharacteristically charitable aside: "A great deal of their difficulty is, of course, the difficulty of the subject and the inapplicability of the scientific method to the subject."[2]

Social science is more difficult than physics? There are aspects of nature to which the scientific method doesn't apply? Surely you're joking, Mr. Feynman.

1. Anonymous, in *Charles Spearman, Wikipedia* (2012).
2. R. P. Feynman, *Surely You're Joking, Mr. Feynman.* (Norton, 1997)

Index

About the Author

Joseph Woelfel is a prominent sociologist and communication scientist, and the author (with Edward L. Fink) of the classic book *The Measurement of Communication Processes: Galileo Theory and Method*. He is the developer of Catpac, a neural network text analysis program widely used around the world. After a review of all the Social Science literature commissioned by the U. S. Army, the Rand Corporation concluded: "In many ways, Woelfel's theory was the closest that any social science approach came to providing the basis for an end-to-end engineering solution for planning, conducting, and assessing the impact of communications on attitudes and behaviors." Woelfel is currently Professor of Communication at the University at Buffalo, State University of New York.

About the Book

The book traces the social network of physical science back to the Ionian Greeks and beyond, while the social network of social scientists has its roots in Athens. The two networks are distinct, and the culture of the Ionian network is based on a comparative measurement model which makes scientific observation possible. The Athenian model of the social scientists is categorical, and does not allow for scientific observation.

Made in the USA
Middletown, DE
18 April 2015